Progress in Soil Science

Aims and Scope

Progress in Soil Science series aims to publish books that contain novel approaches in soil science in its broadest sense – books should focus on true progress in a particular area of the soil science discipline. The scope of the series is to publish books that enhance the understanding of the functioning and diversity of soils in all parts of the globe. The series includes multidisciplinary approaches to soil studies and welcomes contributions of all soil science subdisciplines such as: soil genesis, geography and classification, soil chemistry, soil physics, soil biology, soil mineralogy, soil fertility and plant nutrition, soil and water conservation, pedometrics, digital soil mapping, proximal soil sensing, soils and land use change, global soil change, natural resources and the environment.

More information about this series at http://www.springer.com/series/8746

Alexandre M.J.-C. Wadoux
Brendan Malone • Budiman Minasny
Mario Fajardo • Alex B. McBratney

Soil Spectral Inference with R

Analysing Digital Soil Spectra using the R Programming Environment

Springer

Alexandre M.J.-C. Wadoux
The University of Sydney
Sydney, NSW, Australia

Brendan Malone
CSIRO
Canberra, ACT, Australia

Budiman Minasny
The University of Sydney
Sydney, NSW, Australia

Mario Fajardo
The University of Sydney
Sydney, NSW, Australia

Alex B. McBratney
The University of Sydney
Sydney, NSW, Australia

ISSN 2352-4774 ISSN 2352-4782 (electronic)
Progress in Soil Science
ISBN 978-3-030-64895-4 ISBN 978-3-030-64896-1 (eBook)
https://doi.org/10.1007/978-3-030-64896-1

This Springer imprint is published by the registered company Springer Nature Switzerland AG
The registered company address is: Gewerbestrasse 11, 6330 Cham, Switzerland

Foreword

The soil, a key factor in agricultural production and in particular in agro-ecology, is a phenomenon that is still largely under-measured. It is essential to develop methods to know the soil better, understand how it works, assess its potential, and estimate its state in order to guide corrective actions or estimate the ecosystem services it provides (carbon storage, geochemical cycle closing, etc.). These data will be indispensable in the development of a 'digital agro-ecology', which is largely based on understanding the interactions between all factors of production – soil, climate, species, varieties and their interactions (the basis of agro-ecology!), and other inputs – in order to optimize them. These soil analysis methods must be quantitative, fast, easy to implement, precise, and, if possible, realized with devices that can be easily transported by operators, on-ground machines, or drones for in-field measurements with minimum disturbance and handling of soil samples. Spectral methods meet several of these criteria and can therefore revolutionize the quantitative characterization of soils, for the development and realization of 'digital agro-ecology'. Their main advantage is their ease of implementation with, for some of them, minimal sample preparation. Moreover, while the basic principle is the same for all of these methods (i.e. based on the interaction of electromagnetic radiation with the object of analysis – here the soil), the wide electromagnetic range (UV, visible, NIR, MIR, LIBS, Raman, XRF, etc.) produces various types of digital spectra, which give access to a range of soil properties. Their main drawback is that they are indirect methods which need a calibration – that is, a model – to be constructed between the spectrum and the soil property of interest. Conventionally, these calibrations are built by 'black-box' software, based on algorithms which are generally not known or understood and which are limited in terms of adjustment capacities. Initiatives are emerging to help researchers to better master chemometric techniques in order to build processing pipelines in which they control everything: for example, here in Montpellier, INRAE has launched such an initiative, Chemhouse, led by Jean-Michel Roger. No doubt, there are similar initiatives around the world. Indeed, this book, which focuses on quantitative chemometric methods, is another eagerly awaited response to this need for knowledge-guided spectral analyses to make the most of digital spectra and improve the quality and usefulness of information from soil spectroscopy.

Véronique Bellon-Maurel

Director of #DigitAg, the Digital Agriculture Convergence Laboratory, Montpellier, France
Deputy head of the Mathnum Department, INRAE (mathematics, computer and data sciences, digital technologies)

Preface

Digital spectroscopy is one of the new tools of the state-of-the-art soil scientist. Properly processed spectral data satiate the demand for cheap and accurate soil information required for precision agriculture and food production, earth system modelling, climate change mitigation, and general soil process parametrization. Our understanding of soil spectroscopy has advanced rapidly in the last two decades. The technological developments of cheaper and more accurate sensors coupled with the advent of new numerical tools have contributed to this significant improvement.

The focus of the book is on the techniques of using spectral data for characterizing soil. Spectral data may come from different sensors and wavelengths, γ rays, X rays, and infrared, among others, and from scans made either close to the soil material – in the field or laboratory, or remotely, for example, when the sensor is mounted on a plane or satellite. Most of the examples here are based on the infrared part of the spectrum, largely because of its demonstrated utility for soil science. We present explanation and code in a didactic way that can handle all kinds of spectral data however, in the hope that this book will contribute to the development of common procedures for soil spectral analysis and data sharing whatever the wavelength range. We also hope that this book can be used for developing training courses and capacity building.

Sydney, NSW, Australia
July 2020

Alexandre M.J-C. Wadoux
Brendan Malone
Budiman Minasny
Mario Fajardo
Alex B. McBratney

Acknowledgements

We thank those who have contributed, directly or indirectly, to the development of the materials presented in this book. Leonardo Ramirez-Lopez (BUCHI Labortechnik, Zurich) is pretty much solely responsible for building and maintaining packages for spectral similarity and modelling of complex spectral datasets using local calibration algorithms. Colleagues at the University of Sydney, especially Edward Jones, have given feedback and helped in the development of materials over the years. Finally, we thank also the students and the workshop participants for continuous feedbacks and questions.

Endorsements

Soil Spectral Inference with R offers an introduction and hands-on practical approach on soil spectroscopy for anyone who wants to extend their understanding and capabilities in soil spectroscopy. Since the basics of soil spectral analysis but also new developments are addressed, the book is suitable for beginners and more experienced scientists, either as study material or as a basis for capacity building.

As such, it addresses an important step in the entire soil spectroscopy workflow that stretches from sampling, wet and dry chemistry measurements, quality assessment and control, spectral library development, calibration transfer, spectral data analysis, and use of the resulting soil data for monitoring or mapping of soil properties and functions to data serving. This entire workflow (and best practice guidelines for its parts) is currently addressed in two international initiatives. The initiative to build a Global Soil Spectral Calibration Library and Estimation Service by the Global Soil Laboratory Network (GLOSOLAN), for now mainly focused on mid-infrared lab analysis and prediction including capacity building for both, and the IEEE initiative P4005 Standard Protocol and Scheme for Measuring Soil Spectroscopy, focusing on near-infrared lab analysis. The GLOSOLAN initiative aims to provide a standard global dataset and easy-to-use tools for labs and scientists for spectral data processing. This book can foster understanding of the spectral analysis used and as such can be instrumental to a proper use of the tool or service and help in correct interpretation of the results.

Fenny van Egmond, ISRIC – World Soil Information, Wageningen – the Netherlands, co-lead of the GLOSOLAN initiative for a Global Soil Spectral Calibration Library and Estimation Service

Soil Spectral Inference with R provides a step-by-step description on how soil spectroscopic data can be modelled in the R modelling environment. Starting with a detailed description on soil spectroscopy, this book provides a comprehensive set of tools and techniques used in diffuse reflectance spectroscopy approach for assessing soils. Theoretical concepts and equations are presented in each chapter along with relevant R codes and sample outputs both in the form of data and figures. The references used in this book are up to date and the parts on (a) noise removal, (b) different similarity measures, (c) subsampling approaches, and (d) spectral transformations are going to help students and soil professionals explore new ways and means of analyzing spectral data. With the graphical illustrations of results from almost every segment of example R codes, this book retains its visual presentation style and is expected to serve as a perfect sequel to the earlier book *Using R for Digital Soil Mapping* published from the same group. I highly recommend this book to my students and researchers who are exploring the use of diffuse reflectance spectroscopy for soil analysis.

Bhabani S. Das, Agricultural and Food Engineering Department, Indian Institute of Technology Kharagpur-India

This timely and opportune book will help readers to deal with the increasingly complex kaleidoscope of tools and lines of codes applied to soil spectroscopy. It touches aspects of conversion from wavelength to wavenumber up to the sharing of soil spectral libraries and the need for spectral standardization between laboratories. In this way, the text comes in handy for neophytes and professionals alike. Within its concise contents, the book covers a getting started with R, a list of useful spectroscopy packages, and data handling parts. Yet, it also covers topics from pre-processing to exploratory soil spectral analysis, which will be of interest to competent users, leading towards more proficient levels. Model calibration and estimating soil properties are also implemented and will allow readers to develop their skills up to the expert level with full application of soil spectroscopy. Besides ready-to-use lines of code, datasets with soil spectra and laboratory analytical data were also made available through computer coding. And authors have also compiled all data and functions, used in the book, in a single R package called soilspec, available in the open source software development environment and social network GitHub. Therefore, it's my pleasure to endorse the book Soil Spectral Inference with R, which embodies a complete guidance for lecturing and learning soil spectral inference using the statistical computing environment R.

Alexandre ten Caten, Department of Agriculture, Biodiversity and Forests, Federal University of Santa Catarina-Brazil

Soil Spectral Inference with R can help us to build an integrated application of a soil spectral inference system from scratch using the R platform. This book elaborates on the whole process of soil spectral inference with detailed and practical R codes, from importing and pre-processing of spectra to model calibration and validation. There are many valuable routines in this book, especially for the vis-NIR spectra, such as bagging PLSR for calibration and EPO for the removal of moisture effect. These methods can be examined via the use of example datasets and readily transferred to real-life applications. I, thus, highly recommend this book to anyone who is engaged in the exploration and application of soil spectra.

Changkun Wang, Institute of Soil Science, Chinese Academy of Sciences, Nanjing-China

Contents

About the Authors

Alexandre Wadoux is research associate in soil science at the University of Sydney and member of the Sydney Institute of Agriculture, Australia.

Brendan Malone is a senior research scientist in soil science at CSIRO Canberra, Australia.

Budiman Minasny is professor of soil-landscape modelling at the University of Sydney, Australia.

Mario Fajardo is a postdoctoral research fellow at the Precision Agriculture Laboratory, University of Sydney, Australia.

Alex McBratney is professor of digital agriculture and soil science at the University of Sydney and director of the Sydney Institute of Agriculture, Australia.

Chapter 1
Introduction

Soil provides a multitude of key ecosystem services such as food production, climate change adaptation, nutrient and water cycling and carbon sequestration (Dominati et al. 2010). Ongoing global environmental change has put unprecedented pressure on soil, resulting in significant and widespread degradation and erosion. The soil science community is tasked to deliver timely, nuanced and high-quality thematic soil data and knowledge to assess and monitor soil change (Sanchez et al. 2009). This is reflected by recent initiatives to provide soil information to populate regional, national and worldwide soil databases (Grunwald et al. 2011). Soil data are conventionally acquired through soil surveys coupled with laboratory analyses. The methods to obtain soil information are often impractical because they are expensive, require time-consuming field campaigns and use chemical reagents for soil analysis (McCarty et al. 2002; Brown et al. 2006; Ben-Dor et al. 2009; Stenberg et al. 2010). The use of sensors for characterizing chemical, mineralogical, biological and physical properties of the soil has thus gained lots of traction in soil research. Advances in sensors and software are occurring at a rapid pace. Soil sensing, in particular the use of soil spectroscopy, is now widely available using a range of modalities and wavelengths across the electromagnetic spectrum.

Soil spectroscopy can characterize soil properties efficiently. Soil spectroscopy can be simply defined as the study of the spectral signature of a soil material (Nocita et al. 2015). The spectral signature relates to soil characteristics such as organic and mineral components. Spectroscopic measurements are fast, cost-effective and non-destructive and can be made both in the laboratory and *in situ* in the field. Soil composition and characteristics are encoded in the spectrum at specific wavelengths of the electromagnetic spectrum. For example, mid-infrared spectra have encoded information on soil mineralogy or soil organic matter composition, which can be assessed quantitatively or qualitatively using the absorption or reflectance at specific wavelengths (Viscarra-Rossel et al. 2016).

A. M. J.-C. Wadoux et al., *Soil Spectral Inference with R*, Progress in Soil Science,
https://doi.org/10.1007/978-3-030-64896-1_1

Figure 1.1 summarizes the electromagnetic spectrum over an extensive range of wavelengths and frequencies. The electromagnetic spectrum ranges from the γ rays to radiowaves. The visible (to the human) portion of the electromagnetic spectrum is between 0.4 and 0.75 μm. Radiowaves have long wavelengths that can reach several hundreds of metres, while high-energy γ rays have wavelengths shorter than 10^{-13} m. Each portion of the electromagnetic spectrum relates to specific soil properties or characteristics. For example, the visible part contains information on soil colour, while the γ rays, X rays and infrared spectra are useful to estimate soil properties, especially elemental composition and soil mineralogy.

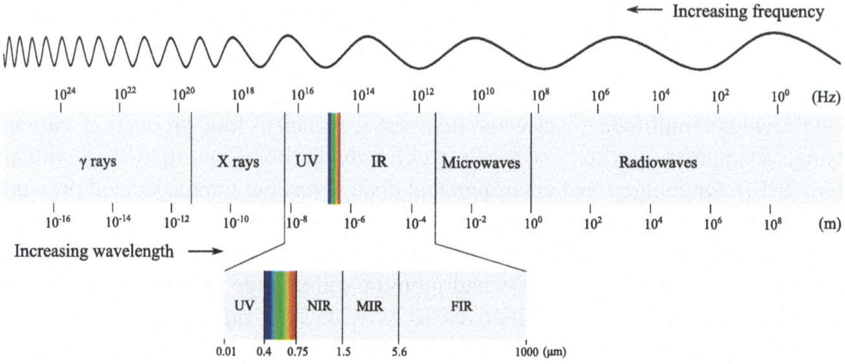

Fig. 1.1 Components of the electromagnetic spectrum. (After Lillesand et al. 2015)

1.1 Spectroscopy in Soil Science

Interest in spectroscopy for soil started as early as the 1920s with studies analysing the mineral (Hendricks and Fry 1930) and later organic composition (e.g. Hunt et al. 1950 or Holmes and Toth 1957) of soils, but it was in the 1970s that scientists started to investigate the direct relationships between soil spectral information and soil properties. Condit (1970), for instance, developed a soil spectral library which quickly became a classical tool for soil scientists (Ben-Dor et al. 2009). The large use of spectral information in soil science was made possible with the advancement of computer and information technology. A big change came in the 1980s and 1990s when spectral instruments were transformed from analogue (chart-recording) to digital devices producing long bivariate data streams of wavelength (or its homologue) and intensity representing the analogue spectra. These data streams are digital spectra usually of length 2^9 to 2^{12}.

Spectroscopy is currently used in a large number of applications: to characterize soil minerals (Viscarra-Rossel and Webster 2012), organic matter (Gerzabek et al. 2006; Ertlen et al. 2010), colour (Viscarra-Rossel and Webster 2011), but also texture (Minasny et al. 2008), iron oxides (Malengreau et al. 1996), carbonates (Grinand et al. 2012), salinity (Nawar et al. 2014) and soil quality indicators

(Cécillon et al. 2009). Soil colloids such as clay minerals are also detected by X ray (~10 nm–10 pm) diffraction (Wilson and Cradwick 1972), by detection of peaks in the pure minerals and comparing them to those recorded on the soil sample. Soil aggregates can be characterized by visible, near and mid-infrared spectroscopy (Cañasveras et al. 2010; Askari et al. 2015). At a larger spatial scale, soil variation and diversity can be characterized by γ rays or microwaves such as radar remote sensing (Cook et al. 1996; Weihermüller et al. 2007).

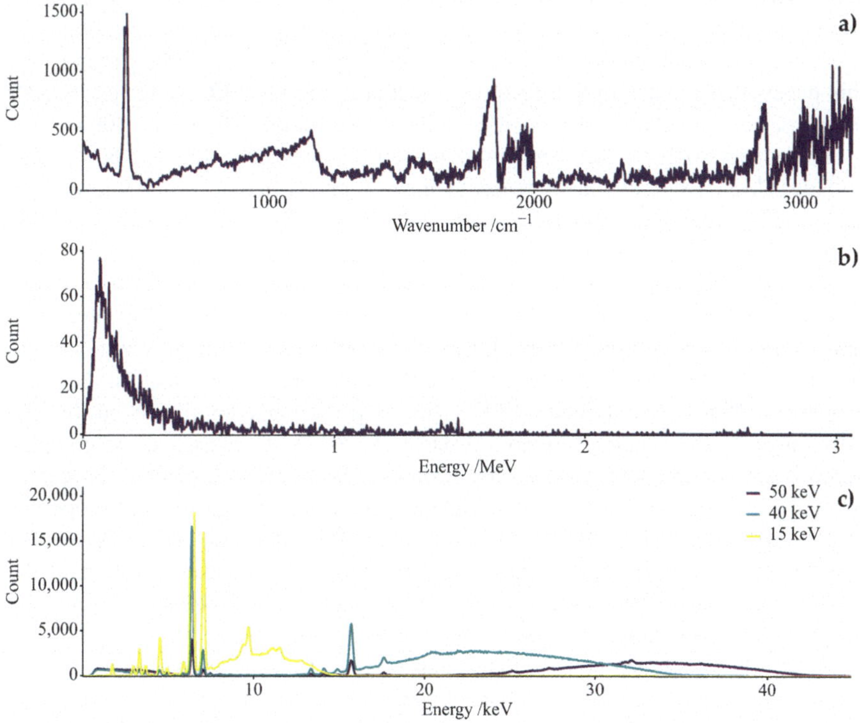

Fig. 1.2 Examples of Raman (**a**), γ rays (**b**) and X ray fluorescence (**c**) spectra of a soil sample from the Geeves (Geeves et al. 1994) dataset. The x-axis unit is in either wavenumber (cm^{-1}), megaelectron volts (Mev) or kiloelectron volts (keV)

Figure 1.2 is an example of three spectra from Raman, γ ray and X ray fluorescence (XRF) spectroscopies. Raman vibrational spectroscopy is useful to characterize soil substances and requires minimal soil sample preparation. γ ray spectroscopy relates to soil mineralogical properties and geochemistry of the soil sample by measuring the natural emission of γ rays and anthropogenic radionuclides, e.g. Caesium-137. An XRF spectrum relates to soil elemental composition.

The visible and infrared range of the electromagnetic spectrum has garnered much interest in soil science. The measurement of the infrared spectrum of soil samples enables the quantification of several soil properties from their spectral response in a faster and cheaper way than by conventional methods of soil analyses

(Stenberg et al. 2010; Bellon-Maurel and McBratney 2011). In addition, recording an infrared spectrum does not make use of any chemical reagents and can be done both in the laboratory or for in-field soil analysis (Ramirez-Lopez et al. 2019). Infrared spectra are sensitive to both organic and inorganic soil materials, making them an excellent tool for quantitative soil assessment. The mid-infrared (MIR) range of the spectrum, in particular, contains more information and direct information on soil organic and mineral components of the soil than the visible and near-infrared (vis-NIR) range. For example, various components of the soil organic matter have very distinct spectral signature in the mid-infrared range. The reason is that the fundamental molecular vibrations occur in the mid-infrared range, while the overtones and combinations occur in the vis-NIR (McCarty et al. 2002). In practice, this means that the absorption features detected in the vis-NIR are fewer, broader and more complex than those recorded in the mid-infrared (Islam et al. 2003).

While spectroscopy has been used in soil science since the 1950s, the last two decades have seen an increase in its use, in particular vis-NIR and MIR spectroscopy, to replace and complement soil analyses. This increase was supported by the development of chemometrics (the application of mathematical and statistical methods to the analysis of chemical data (Varmuza and Filzmoser 2016)), multivariate statistical analysis and the increase in computer resources. Soil properties have complex absorption patterns. Infrared spectral bands are largely non-specific (i.e. they are not linearly related to a single soil property) and overlap between properties (Ben-Dor and Banin 1995). This is particularly significant in the vis-NIR range of the spectra (Soriano-Disla et al. 2014). To extract these complex patterns and obtain quantitative estimates of a soil property, soil scientists have used mathematical transfer functions to correlate spectral wavelengths to soil properties (Viscarra-Rossel et al. 2008). The transfer function is calibrated using the spectral wavelengths as independent variables and the laboratory measured values of the soil properties as the dependent variable. Once calibrated on the spectra, the soil property can be predicted using the spectral information only.

Relatively simple statistical models can be built to transfer the spectra to soil information. Early studies on soil spectroscopy used linear regression models on specific wavelengths. For example, Dalal and Henry (1986) fitted a linear model on three user-defined wavelengths to predict soil moisture, organic carbon and total nitrogen. The large number of wavelengths to consider and the correlation between them made the use of linear models complicated. Techniques for variable selection, such as stepwise variable selection or dimension reduction such as principal component regression, quickly emerged as valuable to handle the multivariate spectral data. A variant of principal component regression called partial least squares regression (PLSR, Abdi 2003) is now routinely used. PLSR relates the soil property values and the principal component scores (a dimension reduction analysis) of the spectra. The PLSR models can handle the full spectra as predictors (not only a few wavelengths) and are not sensitive to the correlation between wavelengths (Janik et al. 2007). In addition, they are substantially faster to calibrate than stepwise linear regression models. In the last two decades, other multivariate analysis techniques have been used, in particular machine learning algorithms. For example, Nawar and Mouazen (2019) used random forest to estimate soil organic carbon on soil samples collected in six fields in the United Kingdom. Viscarra-Rossel and Behrens (2010) compared several linear and non-linear (machine learning) models to calibrate soil spectra on soil properties. Other methods to derive soil information from a spectrum are based on the discrimination on the soil spectral signature such as in absorption feature analysis (Clark and Roush 1984). This book provides implementation to derive soil information from using both multivariate statistical models and absorption feature analysis.

1.2 Populating a Soil Database

The opportunity to retrieve cheap soil information from a spectrum has resulted in the development of soil spectral libraries for the quantification of soil properties at local (Guerrero et al. 2016), regional (Gogé et al. 2012) or global (Viscarra-Rossel et al. 2016) scales. Nowadays, several institutions provide spectral libraries with spectra scanned on pure materials, for example, minerals, vegetation or rocks. The United States Geological Survey (USGS) spectral library version 7 (Kokaly et al. 2017) contains several thousands of spectra of different materials for the ultraviolet to the far infrared (0.2 to 200 microns [μm]). Other libraries are exclusive to soil samples, like the Land Use/Cover Area statistical Survey (LUCAS) compiled in Europe (Orgiazzi et al. 2018). By 2018, this soil spectral library had approximately 45,000 soil samples with spectra in the vis-NIR regions and soil attributes such as pH, organic carbon and cation exchange capacity, among others.

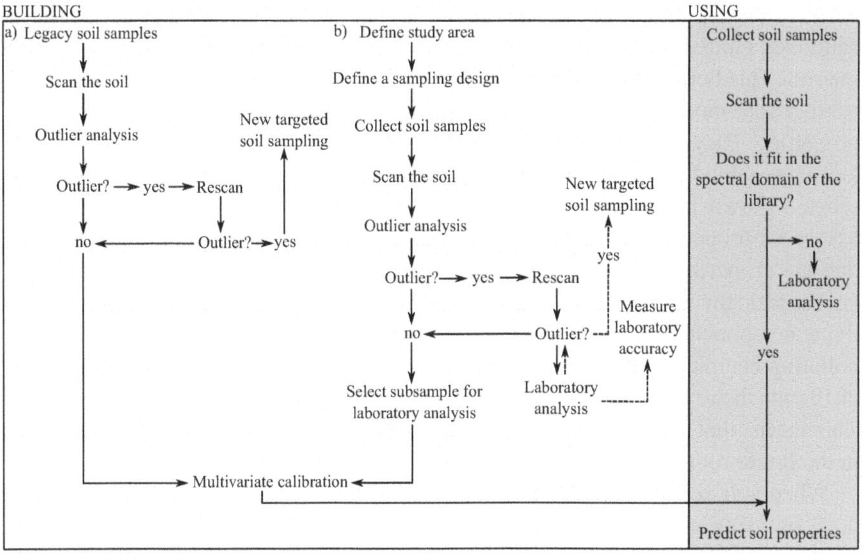

Fig. 1.3 Simplified scheme for building a soil spectral library. (Adapted from Viscarra-Rossel and McBratney 2008). The steps describe how to build a spectral library using (**a**) legacy soil samples or (**b**) a new soil sampling. The scheme explains both the development and the use of the soil spectral library

To build a conventional soil spectral library, Viscarra-Rossel et al. (2008) defined three key requirements: 1. it should contain as many and as representative as possible soil samples to represent the soil spatial variability in the study area, 2. the soil sampling and scanning should be made with caution as all change in the soil sample and scanning procedure is embodied in the spectrum and 3. the laboratory measurement of the soil properties should be accurate. Figure 1.3 illustrates the steps to build a soil spectral library using either legacy soil samples (a) or a new soil sampling (b) and to both build and use the spectral library.

Using legacy soil samples, Fig. 1.3a shows that all soil samples are scanned and the spectra analysed for outliers. If a spectrum is considered as an outlier, the soil is re-scanned. When the re-scanned soil sample is still considered as an outlier, one should consider a new targeted soil sampling for this specific outlier soil sample. After the outlier detection, the spectra are correlated to the values of laboratory-analysed soil properties (e.g. soil organic carbon) using a multivariate statistical model. When, conversely, a spectral library is built using new soil sampling, the sampling design plays a key role. The soil samples are collected in the study area of interest, using a sampling design and for a given sample size. The sample size is decided with budget constraints. The soil samples are then scanned and analysed for outliers. When the spectra contain outliers, they can be re-scanned or analysed in the laboratory. The spectral dataset is subsampled to determine which soil samples are sent to the laboratory for conventional analyses. After this step, the spectra and values of the soil properties are correlated using a multivariate statistical model.

In both cases, when the multivariate statistical model is validated and has sufficiently high accuracy, it can be used for prediction on new soil sample spectra. When new soil samples are scanned and added to the library, they should belong to the same population as the soils in the library. If otherwise, it is likely that the calibrated multivariate models will be inefficient at predicting the soil properties of interest. This book provides implementation for all these steps.

To date, most soil spectral libraries have been built for conventional soil properties. They have been shown a useful means of organizing spectra for small farms or individual fields but also larger, regional areas or continents. Under the scheme shown in Fig. 1.3, it is in principle also possible to build a soil spectral library for properties previously unknown at the time of the soil sampling, provided that the properties are identifiable by spectroscopic techniques. For example, this is the case for microbial biomass carbon (i.e. the carbon contained within the living component of soil organic matter) (Mirzaeitalarposht and Kambouzia 2020), polluting chemicals (Paradelo et al. 2016) or microplastic in soil (Corradini et al. 2019) which are now dynamic research areas but were not considered until recently. This means that spectral libraries and the collection of soil spectra might have use in the future for purposes which we currently disregard.

When soil scientists build a model to link soil properties to the spectra, it is done by using software for statistical analysis. One of the claims made with the availability of spectroscopic measurement devices is the provision of easy-to-follow commercialized software and numerical implementation, which make complicated statistical treatment practicable. While these implementations have provided the majority with tools to produce soil information, they are often used at the expense of a deeper understanding of the techniques required to treat a specific soil spectral library. Useful research has been made in developing functions and code to perform these tasks in open-source software, such as in R (R Core Team 2018). This book will provide a step towards the implementation of spectroscopic analysis techniques and their use in an open, accessible and comprehensible manner and with a view to improving these methods.

1.3 Objectives of This Book

This book is a step-by-step guide to processing soil spectra, particularly from the visible and infrared range of the electromagnetic spectrum. This book is fully reproducible and can serve as a basis for teaching soil spectroscopy to undergraduate students. The examples are implemented in the R programming language, for which the reader is expected to have some basic knowledge. All the data used in the examples, together with the R functions, are provided in a book-associated R package freely accessible online. Instructions to obtain and install the package are provided further on.

Specific topics covered in this book are:

- Importing and plotting spectra in R.
- Pre-processing the spectra.
- Using dimension reduction techniques to visualize the spectra.
- Obtaining soil information (mineralogy, colour) directly from the spectra.
- Outlier detection in the spectral space.
- Similarity measures between spectra.
- Sampling designs and determining the optimal number of soil samples for laboratory analysis.
- Multivariate calibration.
- Soil spectral inference systems.

We also included examples and code for additional (more specific) spectral treatments such as:

- Transferring spectra between instruments.
- Removing the effect of external factors affecting the spectrum, such as soil moisture.

References

Abdi H (2003) Partial least square regression (PLS regression). Encycl Res Methods Soc Sci 6:792–795

Askari MS, Cui J, O'Rourke SM, Holden NM (2015) Evaluation of soil structural quality using VIS–NIR spectra. Soil Tillage Res 146:108–117

Bellon-Maurel V, McBratney AB (2011) Near-infrared (NIR) and mid-infrared (MIR) spectroscopic techniques for assessing the amount of carbon stock in soils–Critical review and research perspectives. Soil Biol Biochem 43:1398–1410
Ben-Dor E, Banin A (1995) Near infrared analysis (NIRA) as a method to simultaneously evaluate spectral featureless constituents in soils. Soil Sci 159:259–270
Ben-Dor E, Chabrillat S, Demattê JAM, Taylor GR, Hill J, Whiting ML, Sommer S (2009) Using imaging spectroscopy to study soil properties. Remote Sens Environ 113:S38–S55
Brown DJ, Shepherd KD, Walsh MG, Mays MD, Reinsch TG (2006) Global soil characterization with VNIR diffuse reflectance spectroscopy. Geoderma 132:273–290
Cañasveras JC, Barrón V, Del Campillo MC, Torrent J, Gómez JA (2010) Estimation of aggregate stability indices in Mediterranean soils by diffuse reflectance spectroscopy. Geoderma 158:78–84
Cécillon L, Barthès BG, Gomez C, Ertlen D, Génot V, Hedde M, Stevens A, Brun J-J (2009) Assessment and monitoring of soil quality using near-infrared reflectance spectroscopy (NIRS). Eur J Soil Sci 60:770–784
Clark RN, Roush TL (1984) Reflectance spectroscopy: quantitative analysis techniques for remote sensing applications. J Geophys Res Solid Earth 89:6329–6340
Condit HR (1970) The spectral reflectance of American soils. Photogramm Eng 36:955–966
Cook SE, Corner RJ, Groves PR, Grealish GJ (1996) Use of airborne gamma radiometric data for soil mapping. Soil Res 34:183–194
Corradini F, Bartholomeus H, Lwanga EH, Gertsen H, Geissen V (2019) Predicting soil microplastic concentration using vis-NIR spectroscopy. Sci Total Environ 650:922–932
Dalal RC, Henry RJ (1986) Simultaneous determination of moisture, organic carbon, and total nitrogen by near infrared reflectance spectrophotometry. Soil Sci Soc Am J 50:120–123
Dominati E, Patterson M, Mackay A (2010) A framework for classifying and quantifying the natural capital and ecosystem services of soils. Ecol Econ 69:1858–1868
Ertlen D, Schwartz D, Trautmann M, Webster R, Brunet D (2010) Discriminating between organic matter in soil from grass and forest by near-infrared spectroscopy. Eur J Soil Sci 61:207–216
Geeves GW, Cresswell HP, Murphy BW, Gessler PI, Chartres CJ, Little IP, Bowman GM (1994) Physical, chemical and morphological properties of soils in the wheat-belt of southern NSW and northern Victoria. NSW Department of Conservation; Land Management/CSIRO Division of Soils Occasional Report, CSIRO
Gerzabek MH, Antil RS, Kögel-Knabner I, Knicker H, Kirchmann H, Haberhauer G (2006) How are soil use and management reflected by soil organic matter characteristics: a spectroscopic approach. Eur J Soil Sci 57:485–494
Gogé F, Joffre R, Jolivet C, Ross I, Ranjard L (2012) Optimization criteria in sample selection step of local regression for quantitative analysis of large soil NIRS database. Chemom Intell Lab Syst 110:168–176
Grinand C, Barthes BG, Brunet D, Kouakoua E, Arrouays D, Jolivet C, Caria G, Bernoux M (2012) Prediction of soil organic and inorganic carbon contents at a national scale (France) using mid-infrared reflectance spectroscopy (MIRS). Eur J Soil Sci 63:141–151
Grunwald S, Thompson JA, Boettinger JL (2011) Digital soil mapping and modeling at continental scales: finding solutions for global issues. Soil Sci Soc Am J 75:1201–1213
Guerrero C, Wetterlind J, Stenberg B, Mouazen AM, Gabarrón-Galeote MA, Ruiz-Sinoga JD, Zornoza R, Viscarra-Rossel RA (2016) Do we really need large spectral libraries for local scale SOC assessment with NIR spectroscopy? Soil Tillage Res 155:501–509
Hendricks SB, Fry WH (1930) The results of X-ray and microscopical examinations of soil colloids. Soil Sci Soc Am J 11:194–195
Holmes RM, Toth SJ (1957) Physico-chemical behavior of clay-conditioner complexes. Soil Sci 84:479–488
Hunt JM, Wisherd MP, Bonham LC (1950) Infrared absorption spectra of minerals and other inorganic compounds. Anal Chem 22:1478–1497
Islam K, Singh B, McBratney AB (2003) Simultaneous estimation of several soil properties by ultra-violet, visible, and near-infrared reflectance spectroscopy. Soil Res 41:1101–1114

Janik LJ, Skjemstad J, Shepherd K, Spouncer L (2007) The prediction of soil carbon fractions using mid-infrared-partial least square analysis. Soil Res 45:73–81

Kokaly RF, Clark RN, Swayze GA, Livo KE, Hoefen TM, Pearson NC, Wise RA, Benzel WM, Lowers HA, Driscoll RL, others (2017) USGS spectral library version 7. US Geological Survey

Lillesand T, Kiefer RW, Chipman J (2015) Remote sensing and image interpretation. Wiley, New York

Malengreau N, Bedidi A, Muller J-P, Herbillon AJ (1996) Spectroscopic control of iron oxide dissolution in two ferralitic soils. Eur J Soil Sci 47:13–20

McCarty GW, Reeves JB, Reeves VB, Follett RF, Kimble JM (2002) Mid-infrared and near-infrared diffuse reflectance spectroscopy for soil carbon measurement. Soil Sci Soc Am J 66:640–646

Minasny B, McBratney AB, Tranter G, Murphy BW (2008) Using soil knowledge for the evaluation of mid-infrared diffuse reflectance spectroscopy for predicting soil physical and mechanical properties. Eur J Soil Sci 59:960–971

Mirzaeitalarposht R, Kambouzia J (2020) Development of mid-infrared spectroscopic feature-based indices to quantify soil carbon fractions. Eurasian Soil Sci 53:73–81

Nawar S, Buddenbaum H, Hill J, Kozak J (2014) Modeling and mapping of soil salinity with reflectance spectroscopy and landsat data using two quantitative methods (PLSR and MARS). Remote Sens 6:10813–10834

Nawar S, Mouazen AM (2019) On-line vis-NIR spectroscopy prediction of soil organic carbon using machine learning. Soil Tillage Res 190:120–127

Nocita M, Stevens A, van Wesemael B, Aitkenhead M, Bachmann M, Barthès B, Dor EB, Brown DJ, Clairotte M, Csorba A, others (2015) Soil spectroscopy: an alternative to wet chemistry for soil monitoring. In: Advances in agronomy. Elsevier, Burlington, pp 139–159

Orgiazzi A, Ballabio C, Panagos P, Jones A, Fernandez-Ugalde O (2018) LUCAS soil, the largest expandable soil dataset for Europe: a review. Eur J Soil Sci 69:140–153

Paradelo M, Hermansen C, Knadel M, Moldrup P, Greve MH, Jonge LW de (2016) Field-scale predictions of soil contaminant sorption using visible–near infrared spectroscopy. J Near Infrared Spectrosc 24:281–291

Ramirez-Lopez L, Wadoux AMJ-C, Franceschini MHD, Terra FS, Marques KPP, Sayão VM, Demattê JAM (2019) Robust soil mapping at the farm scale with vis–NIR spectroscopy. Eur J Soil Sci 70:378–393

R Core Team (2018) R: a language and environment for statistical computing. R Foundation for Statistical Computing, Vienna, Austria

Sanchez PA, Ahamed S, Carré F, Hartemink AE, Hempel J, Huising J, Lagacherie P, McBratney AB, McKenzie NJ, Lourdes Mendonça-Santos M de, others (2009) Digital soil map of the world. Science 325:680–681

Soriano-Disla JM, Janik LJ, Viscarra-Rossel RA, Macdonald LM, McLaughlin MJ (2014) The performance of visible, near-, and mid-infrared reflectance spectroscopy for prediction of soil physical, chemical, and biological properties. Appl Spectrosc Rev 49:139–186

Stenberg B, Viscarra-Rossel RA, Mouazen AM, Wetterlind J (2010) Visible and near infrared spectroscopy in soil science. In: Advances in agronomy. Elsevier, Burlington, pp 163–215

Varmuza K, Filzmoser P (2016) Introduction to multivariate statistical analysis in chemometrics. CRC press, Boca Raton

Viscarra-Rossel RA, Behrens T (2010) Using data mining to model and interpret soil diffuse reflectance spectra. Geoderma 158:46–54

Viscarra-Rossel RA, Behrens T, Ben-Dor E, Brown D, Demattê J, Shepherd KD, Shi Z, Stenberg B, Stevens A, Adamchuk V, others (2016) A global spectral library to characterize the world's soil. Earth-Sci Rev 155:198–230

Viscarra-Rossel RA, Jeon YS, Odeh IOA, McBratney AB (2008) Using a legacy soil sample to develop a mid-IR spectral library. Soil Res 46:1–16

Viscarra-Rossel RA, McBratney AB (2008) Diffuse reflectance spectroscopy as a tool for digital soil mapping. In: Digital soil mapping with limited data. Springer, Berlin, pp 165–172

Viscarra-Rossel RA, Webster R (2012) Predicting soil properties from the Australian soil visible–near infrared spectroscopic database. Eur J Soil Sci 63:848–860
Viscarra-Rossel RA, Webster R (2011) Discrimination of Australian soil horizons and classes from their visible–near infrared spectra. Eur J Soil Sci 62:637–647
Weihermüller L, Huisman JA, Lambot S, Herbst M, Vereecken H (2007) Mapping the spatial variation of soil water content at the field scale with different ground penetrating radar techniques. J Hydrol 340:205–216
Wilson MJ, Cradwick PD (1972) Occurrence of interstratified kaolinite-montmorillonite in some Scottish soils. Clay Miner 9:435–437

Chapter 2
Getting Started with R

R provides a convenient and flexible data-analytic environment for soil spectral data. R is a programming language and a software facility for data manipulation, statistical analysis and graphics. R is an implementation of the S language developed at Bell Laboratories (Venables et al. 2009) in the 1980s. While R is an integrated environment for data manipulation, it is mostly used for statistical analyses. R builds is a so-called 'GNU' project, i.e. it is public domain and all resources are freely accessible, unlike some other programming languages such as Matlab.

The development of R for statistical analyses relies on the users who develop and maintain a large variety of packages. A few of them are built into the R-base system, but all existing packages are accessible online (see Sect. 2.6). One of the main advantages of R is the amount of information and resources that any user can find on the Internet and the constant evolution of resources.

This chapter is a very short introduction to the use of R and to one of the user-friendly graphical user interfaces (GUI) called RStudio. This chapter provides explanations of the main commands, programming tools and graphical functions needed to understand and follow the content of the book. This chapter is written for users with little or without any previous programming experience but is not enough by itself to be proficient in programming using R. In the latter case, the reader is redirected to Sect. 2.6 or to Venables et al. (2009) for further references.

2.1 Use of R and RStudio

Installing R Installing the latest version of R is freely and legally accessible from the Comprehensive R Archive Network (CRAN) website https://cloud.r-project.org/ with the following steps.

1. Click on *Download R for Windows* assuming you work on Windows; otherwise, select the platform Linux or (Mac) OS X.

A. M. J.-C. Wadoux et al., *Soil Spectral Inference with R*, Progress in Soil Science,
https://doi.org/10.1007/978-3-030-64896-1_2

2. Click on *base* or *install R for the first time* which redirects to the base package page for the latest version of R available.
3. Click on *Download R 3.6.2 for Windows* (Note that at the moment of writing, version 3.6.2 is the latest) and save the executable file.
4. Locate and click on the executable file. Select the default answers for all questions. R installs itself automatically. You should see an R icon on your desktop. If you click on this icon, R will open as a command windows, and you can start using it. Most users, however, will find it hard to use R in this way as it does not have a GUI. Many freeware GUI are available for R. In this book, we recommend to use one of the most common, called RStudio.

Installing RStudio RStudio is an interface developed to improve the R user experience. RStudio builds the interface on the background R installation and includes some core functionalities such as visualization and code editor panels and of course the R console. Installing the latest version of RStudio is freely accessible on the RStudio website with the following steps:

1. Go to https://rstudio.com/products/rstudio/ and click on *RStudio Desktop*.
2. Choose the Open Source Edition (free) of RStudio and click on *Download RStudio Desktop*.
3. Click on *Download* under the RStudio Desktop icon.
4. Choose the correct platform. Assuming you work on Windows, click on *RStudio-1.2.5033.exe* (Note that at the moment of writing, the version 1.2 is the latest), and save the executable file on your computer.
5. Locate and click on the executable file. Select the default answers for all questions. RStudio installs itself automatically. RStudio should now be installed in your computer. Click on the RStudio icon to open the interface presented in Fig. 2.1.

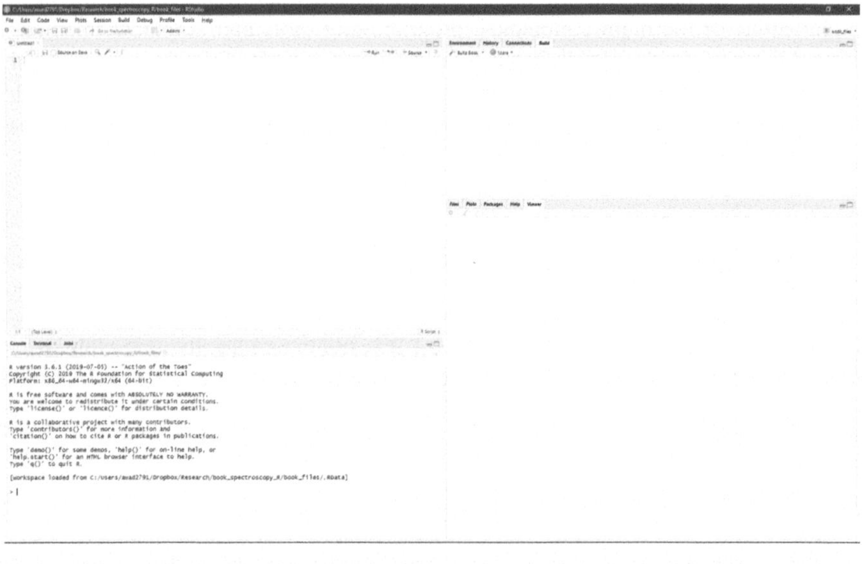

Fig. 2.1 The RStudio interface with the four windows. The upper left window is hidden by default but can be opened by clicking the file menu, then New File and then R script

The RStudio interface is composed of four windows (Fig. 2.1) called *Source* (upper left window), *Environment and history* (upper right window), *Console* (bottom left window) and *Files, Plots, Packages, Help* (bottom right window). R code is executed in the console window. When typing commands in the console window, the output is printed directly. The *Source* window is a built-in text editor and enables the user to edit and save the R scripts. Writing an R script in this window does not run it. For this you need to click *Run* for R to execute the command. In the *Environment and history* window, you can see the data and variables that RStudio has in the memory (loaded into R), and you can click on them to display the values they contain. The history tab tells you what has already been executed. Finally, the *Files, Plots, Packages, Help* window provides display of the plots, location of the files as well as help and loading facilities for R packages.

R packages The R `base` package contains the basic functions which lets R function as a language: arithmetic, input/output, basic programming support, etc. When one downloads R for the first time, in addition to `base`, a number of other packages are also installed that add to the core R functionality and collectively provide an extensive suite of functions that enables generic data analysis tasks. The great thing about R, however, is that it is also very much contributor driven. It is the collective of user contributions via functions and packages development that have made R such a rich computing environment for any manner data analysis tasks. And this extended functionality continues to grow where to date (viz. June 2020) there are over 15,000 packages available from CRAN.

If you have developed a collection of functions and packaged them up, it is possible to share the package on CRAN so that all other R users can also use them. Alternatively, it is increasingly becoming popular for people to share their R packages via code repositories such as R-forge, GitHub and Bitbucket. Not only do these repository services provide a platform to evolve your code in a collaborative project-based way, but some also provide tools to test and compile your packages and to ensure they meet quality specifications to ensure others can use the functions without major difficulties. The advantage of getting your R packages into CRAN however is that they will be easier to find during Internet searches and consequently will get more exposure and a greater number of users. CRAN does substantive testing of the packages prior to publishing. So a fair amount of code improvement is often required after the initial creation to meet the standards set by CRAN. Such standards are mainly around code integrity, documentation clarity and establishing package dependencies.

For packages that are stored and listed in the CRAN website, you can display in R the available packages to download by typing `available.packages()`. Alternatively, you can browse packages by topic in the official CRAN website https://cran.r-project.org/web/views/. The packages already installed in your R session default library can be displayed by using the function `library()`.

In most cases, one is interested in a specific function available in a package. To use this function, you would need to:

1. Install the package, using `install.packages()` and by typing the name of you package inside this function. For example, you will type `install.packages("soilspec")` to install the package associated with this book to your personal library.
2. Load the package into your current R session. It is not sufficient to install the package into your library; you must then load it using the function `library()` or `require()`. For example, you can type `library(soilspec)` in your R console to load the `soilspec` package.
3. The functions are now accessible in your current R session. You can access the functions by typing the name of the function in your console directly. When a function has the same name as existing ones, you can access the function you wish from the specific loaded package by typing the operator `::` after the package name and before the package function. For example, you can type `soilspec::epo` to access the `epo` function from the `soilspec` package.

2.2 Simple Manipulations

Calculator R can be used as a calculator by entering the expression to evaluate in the console and executing. For example, by typing:

```
19 + 90
```

R will return the following:

```
## [1] 109
```

where `+` is used for addition, `*` for multiplication, `-` for subtraction and `^` for exponentiation. As conventionally accepted, the order of doing the arithmetic operations is left to right, and parentheses can be used to force the priority of the operations.

Vector and matrices R is organized as many programming languages upon the organization of numbers into scalars (a single number), vectors (a series of numbers) and matrices (several series of numbers).

A vector is created by using the function `c()`. In the following example, we store the vector into a variable name. This is useful to make operation on the vector or matrix in a later step. To assign a vector or matrix to a variable name, we use the symbols `<-` which is a combination of `<` (less than) and `-` (minus or hyphen). Let us assign a vector or two numbers into the variable name `a1`.

```
a1 <- c(19, 90)
```

The content of the vector can be retrieved simply by typing the variable name into the console.

```
al
```

```
## [1] 19 90
```

The creation of a matrix is explained later in this chapter.

Functions Calculations on vectors or matrices are time-consuming. For example, if you like to take the sum of all numbers contained in the vector `al`, you can use R as a calculator and type:

```
al[1] + al[2]
```

which will return the sum of the first and second numbers contained in `al`. Here the double bracket `[]` is used to access a value of the `al` vector. This operation is impossible for very large vectors, and one should rather use functions instead. Several basic functions are comprised in the `base` R package, and many others are available in contributed packages. You can also create your own functions. A core function is `sum()` which, as the name suggests, sums all numbers contained in a vector such as `al` by:

```
sum(al)
```

Within the brackets of `sum()` are specified the arguments. In this case, the function `sum()` takes a single compulsory argument, that is, a vector of numbers.

Plots R offers a diverse suite of functionality to make graphs. The `base` package has plotting options which are sufficient in most cases. Let us, for example, define a vector of ten numbers, which we call `al2`.

```
al2 <- c(0.24, -0.43, 0.80, -0.39, -0.49, -0.77, 0.18,
         -1.17, 1.43, -1.71)
```

We can plot the vector `al2` simply by using the function `plot()`, for example:

```
plot(al2)
```

which will return on the x-axis the index (position in the vector) of each value and on the y-axis the values of the number for this index. More information on graphs is provided in Sect. 2.5.

2.3 Data Structure

The data structures that follow are routinely used in R and also throughout this book. It is therefore suggested to understand the basic differences between them and their potential applications.

Vectors Vectors are one of the most important types of objects in R; we introduce several other ways to combine or create them. Let us come back to the vector of two numbers called `al2` and use it in combination with another vector of three numbers called `at`.

```
at <- c(2, 41, 5)
```

We can combine the two vectors using the same combine `c()` function than before.

```
newVec <- c(al2, at)
```

where each number in the new vector `newVec` accessed via the command `[]`, for example, using `newVec[3]` accesses the third number in the vector.

To create a vector, one can rely on other functions. We introduce two of them. The first is the `seq()` (sequence) function which creates a sequence of numbers stored in a vector. The second is the `rnorm()` function which creates a sample of values from the normal distribution. The script below shows the use of both of them, which return a vector as output. We first create a sequence of number between 1 and 10 by interval of 1.

```
vec <- seq(1, 10, by = 1)
vec
```

```
##  [1]  1  2  3  4  5  6  7  8  9 10
```

The `vec` vector returns a sequence of ten numbers. In the second case, we sample ten values from the normal distribution. The normal distribution is characterized by two parameters, its mean and standard deviation. We can specify them using the `mean` and `sd` arguments. If we do not specify them, they are kept to their default value of 0 and 1, respectively. Let us sample ten numbers from the normal distribution with mean 0 and standard deviation of 1.

```
vec2 <- rnorm(10)
vec2
```

```
##  [1] -0.6208766 -0.9332151 -1.7487639 -0.5684577  1.1218191  0.9277536
##  [7] -1.5258586  1.9416747  1.0649083  0.9723708
```

The vector printed above shows ten numbers sampled from the normal distribution. We will use it later in this chapter to provide an example of plotting.

Factors Factors are characters which provide a way to handle categorical data. When dealing with categorical variables, it is necessary to ensure that the data are stored as factors. The `factor` function is used to create factors. Both numeric data and characters can be converted to factors. Let us take an example with a vector of seven sampling locations which comprise three possible soil texture classes, called types 1, 2 and 3. Since soil classes are categorical variables, we convert the numeric values to factor.

```
# create a vector of numbers
soilText <- c(1, 2, 3, 3, 2, 3, 1)

# convert the numerics to factors
soilTextF <- factor(soilText)
soilTextF
```

```
## [1] 1 2 3 3 2 3 1
## Levels: 1 2 3
```

Printing the `soilTextF` variable name shows three `levels` which are, in this case, the three possible soil texture classes. This might be useful, but we may want to make our dataset easier to understand and name the soil texture classes according their real classification, which in our case is according to the French 'Aisne' classification (Moeys 2018). The three classes are `AS`, `LSA` and `LAS`.

```
levels(soilTextF) <- c("AS", "LSA", "LAS")
soilTextF
```

```
## [1] AS  LSA LAS LAS LSA LAS AS
## Levels: AS LSA LAS
```

Matrices Matrices are multi-dimensional generalizations of vectors. Matrices can be seen as a table with vectors in their rows or columns. Matrices contain numeric values. Let us create a square matrix with dimensions of two rows and two columns.

```
sqMat <- matrix(data = c(5, 21, 15, 65), nrow = 2, ncol = 2)
sqMat
```

```
##      [,1] [,2]
## [1,]    5   15
## [2,]   21   65
```

The argument `data` specifies the elements of the matrix. Arguments `nrow` and `ncol` specify the number of row and columns of the matrix. We can access specific values of the matrix, for example, by `sqMat[1, 2]` to access the value of the first row, second column, or by `sqMat[,1]` to access all values of the first column. This type of selection is called indexing.

Data frames Data frames are matrix-like structures. A substantial difference between matrices and data frames is that data frames permit both numeric and categorical variables. Data frames can also have column and row names. This is convenient for many purposes. Let us create a `dataframe` object with two columns and two rows. The first column is numeric and called `clay`, the second is categorical and called `class`.

```
datFr <- data.frame(clay = c(25, 12, 56), class = c("AS", "LSA", "LAS"))
datFr
```

```
##   clay class
## 1   25    AS
## 2   12   LSA
## 3   56   LAS
```

One can access specific value of the data frame in the same way as for matrices. A specific column can be accessed by typing the operator `$`, for example, by `datFr$clay`.

Lists A convenient way to store irregular data is to make use of list objects. A list is a vector-like object which takes as input all types of objects. Let us create a list of size 2, i.e. with two elements. The first is a vector, and the second is the data frame called `datFr`.

```
# create a vector of size 3
clay <- c(25, 12, 56)

# make the list
lisFr <- list(clay = clay, datFr = datFr)

# print the list
lisFr
```

```
## $clay
## [1] 25 12 56
##
## $datFr
##   clay class
## 1   25    AS
## 2   12   LSA
## 3   56   LAS
```

Each element in the list can be accessed with the operator `[[]]`, for example, to access the first element of `lisFr`, one can use the command `lisFr[[1]]`. Alternatively, the name of the element (if it has one) can be used, for example, `lisFr$clay`, to access the element called `clay`.

2.4 Programming Tools

This section describes some useful programming tools used in this book. Several of the following programming statements are routinely used and correspond to basic programming skills.

Selecting and subsetting data To manipulate data frames, the most convenient way is to use the bracket notation `[row, column]` or to select by the specific column name. In some cases, however, one would like more advanced options to select or subset data from a data frame. This is the case, for example, when one wants to perform an analysis (e.g. the sum of some properties) for a given category in the data frame. Let us provide an example of subsetting a data frame that we use in this book. We first create a data frame with two columns, one numeric (the content in percent of the soil clay) and another categorical (the soil texture class).

```
datFr <- data.frame(clay = c(25, 12, 56, 22, 6, 31, 45, 65),
                    class = c("LAS", "AS", "LSA", "LSA", "LAS","AS", "AS", "LAS"))
datFr
```

```
##   clay class
## 1   25   LAS
## 2   12    AS
## 3   56   LSA
## 4   22   LSA
## 5    6   LAS
## 6   31    AS
## 7   45    AS
## 8   65   LAS
```

Now we would like to subset the data frame to perform an operation on the clay values larger or equal to 45%. We can apply the following command.

```
datFr[datFr$clay >= 45, ]
```

```
##   clay class
## 3   56   LSA
## 7   45    AS
## 8   65   LAS
```

Now we see that three rows are selected. The command >= means ‘equal or greater than’. We can repeat the same operation by selecting the clay values for a given soil texture class.

```
datFr[datFr$class == "LAS", ]
```

```
##   clay class
## 1   25   LAS
## 5    6   LAS
## 8   65   LAS
```

In this case, we use the operator == which means 'is exactly equal to'. This type of operation can be applied also to columns (instead of rows). In this case, we place the `,` at the beginning of the brackets.

For-loop For-loops are routinely used in programming to repeat a sequence of instructions. Instead of printing manually some commands, we can iterate over all commands to make the operation faster. You need to specify what has to be done and for how many iterations. For example, we want to create a list containing ten of the same object. The object is the matrix called `datFr`. Instead of creating a list using `list()` and adding manually the objects using `list[[1]] <- datFr`, `list[[2]] <- datFr`, etc., we create a for-loop.

```
# create an empty list
listAll <- list()

# make the for loop for each i from 1 to 10
for(i in 1:10){
  # add the object to the list
  listAll[[i]] <- datFr
}

# print the length of the list
length(listAll)
```

```
## [1] 10
```

The object `listAll` contains ten elements, each containing a matrix. The function `length()` returns the length of an object, in our case the number of elements in the list.

Writing a function A large set of functions are available in R packages. Some basic functions are also available in the `base` package. In some cases, one also wants to create customized functions to repeat a specific set of tasks or to simplify large R scripts. In this book, we provide many bespoke functions that are necessary for performing custom workflows for analysis of soil and spectroscopic data. To create a function, it is necessary to specify the input arguments. Let us, for example, create a function that raises to the power of `x` a vector. We create a function called `pow()` with two arguments to specify. The first is `y`, the numeric vector, and the second is `x`, the power. To create a function, we need to use `function()`.

```
pow <- function(y, x){
  output <- y^x
  return(output)
}
```

Our `pow()` function returns an object called `output` that is the numeric vector with same length as the input vector `y`. Note that inside a function, we must provide the `return()` function to print the output. We can now apply our function to a numeric vector. We take a power 6 of a vector of size 4.

```
# create a vector of length 4
vec <- c(2, 1, 3, 0)

# apply our pow() function
pow(y = vec, x = 6)
```

```
## [1]  64   1 729   0
```

2.5 Plotting

Plotting data and making graphs is very easy in R. The function `plot()` is generic (the type of plot depends on the class of the first argument) and applies to many packages. Plotting can also be made using several additional packages such as `ggplot2` but is out of the scope of the book. In this book, we use the `plot()`, the `matplot()` and the `histogram()` functions.

Plot In the simplest case, we plot a numeric vector `x` using `plot(x)`. In this case, it produces a time-series plot, i.e. it produces a plot of the index on the x-axis and the value on the y-axis. In this book, we are rather interested in doing scatterplots. For example, we like to know how far from the 1:1 line are the predicted against observed values. We can do so by using `plot(x, y)` where `x` is a numeric vector of observed and `y` is a numeric vector of predicted values. We create two vectors to produce a scatterplot of `x` against `y`.

```
# vector of the 'observed' values
x <- seq(1, 100, by = 0.5)

# vector of predicted values
y <- x + rnorm(length(x), mean = 0, sd = 5)

# scatterplot x against y
plot(x, y)
```

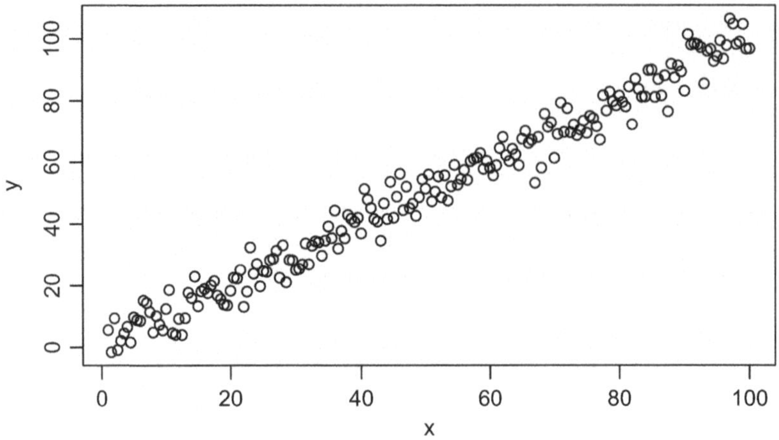

Fig. 2.2 Scatterplot of x against y using the `plot()` function

Figure 2.2 shows what the basic plot function displays. One can make this plot look much better using additional arguments that are set to defaults in Fig. 2.2. The `main` argument provides a title; the `xlab` and `ylab` arguments provide the x-axis and y-axis names, respectively. The `col` argument is the colour and `type` is the way of displaying the observations (e.g. lines or dots). If `type="p"`, i.e. to display dots, the argument `pch` can also be specified as the type of points that one wants. We now improve the figure by setting these additional arguments (Fig. 2.3).

```
plot(x, y,
     main = "Example plot",
     xlab = "Observed /%",
     ylab = "Predicted /%",
     type = "p",
     pch = 16,
     col = rgb(red = 0.5, green = 0.5, blue = 0.5, alpha = 0.7))
```

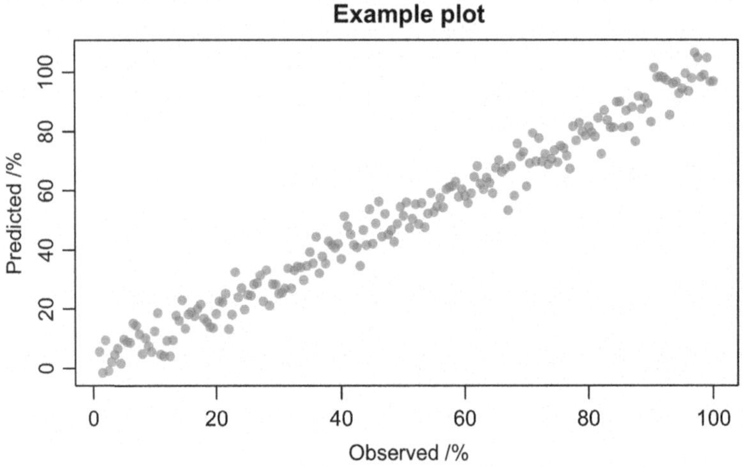

Fig. 2.3 Scatterplot of x against y using the `plot()` function and additional arguments

Matplot The `matplot()` function is similar to `plot()` but plots the columns of a matrix against the columns of another. This is convenient in infrared spectroscopy where one wants to plot the wavelength (a numeric vector) against a matrix containing the spectra. The `matplot()` function has the same additional arguments than the `plot()` function. We take an example plotting the vector of the observed values against realizations of the predictions, in this case three.

```
# make three realizations of the predictions
y1 <- x + rnorm(length(x), mean = 0, sd = 2)
y2 <- x + rnorm(length(x), mean = 0, sd = 4)
y3 <- x + rnorm(length(x), mean = 0, sd = 6)

yAll <- cbind(y1, y2, y3)

# plot using matplot()
matplot(x, yAll,
        xlab = "Observed /%",
        ylab = "Predicted /%",
        type = "p",
        pch = 16)
```

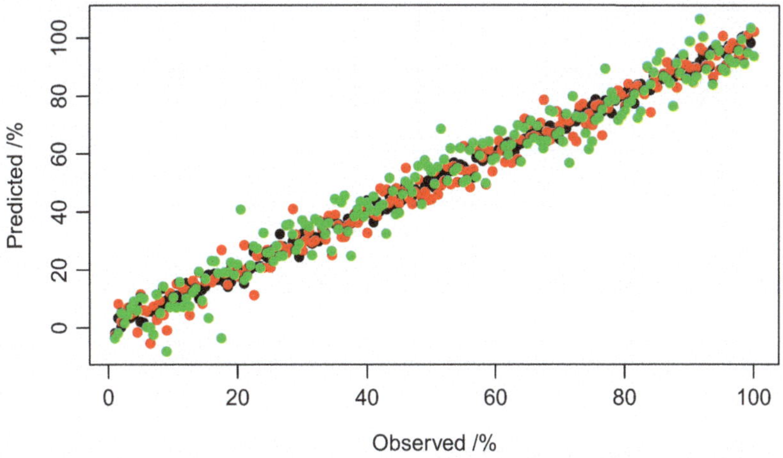

Fig. 2.4 Scatterplot of x against three realizations of y using the `matplot()` function and additional arguments

Figure 2.4 shows the three realizations of `y` plotted against `x`. Each `y` has different colours. The colours can be specified by adding a `col` argument and by adding a vector of colour equal to the number of `y` variables.

Another type of figure that we use in this book is the histogram. The histogram represents the range of the data against the frequency of the data for each specific interval. The `hist()` function can be used for this purpose. Like the `plot()` function, additional arguments can be provided to complement the default plotting options. We can make a histogram of a vector of numeric values sampled from a normal distribution with mean 0 and standard deviation of 1 (Fig. 2.5).

```
# create a vector of size 100
vec <- rnorm(100)

hist(vec,
     main = "Example histogram",
     xlab = "Data range")
```

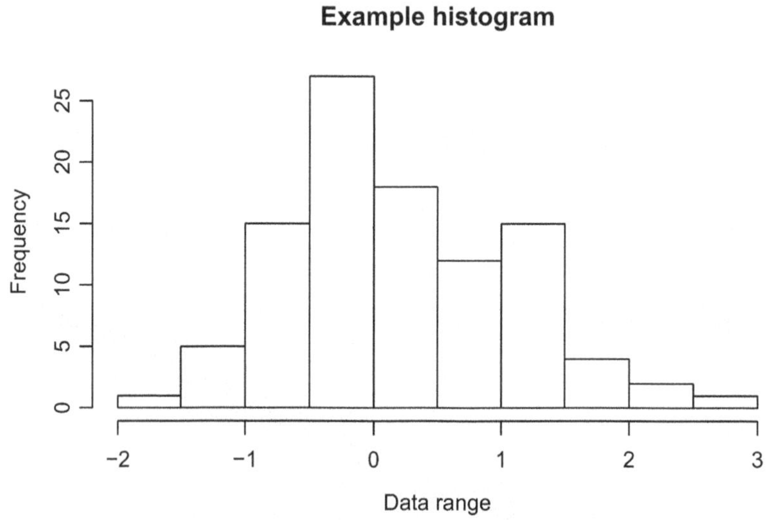

Fig. 2.5 Example histogram from a sample of size 100 of the normal distribution

2.6 Documentation and Help

It is common to ask for help when writing or implementing R codes. One of the great advantages of R is the large network of people using it and providing books, tutorials, examples and forum to answer questions. Within each package, some help is provided. The documentation related to a specific function is attained by typing `help()` in the console, for example, `help(rnorm)` provides help on the `rnorm` function by displaying the default package information on the function. In the package information, each argument of the function is describe along with the object requirement (e.g. it must be a matrix or data frame). For specific packages, the authors may provide a vignette, which are guided scripts with extended descriptions and examples to use the functions and data of the package. To obtain a list of available vignettes, you can type `vignette()` into the console. Vignettes are open from R using the operator `??` in front of a package name and then clicking the vignette if available.

More general resources are found online, for example, by typing a description of the problem in a search engine. It is very likely that someone already found an answer to your question.

For documentation, you can find the official R manuals and other specific documentation under https://cran.r-project.org/other-docs.html or Quick-R. The R community is also very active on forum, and after searching that your question has not yet been answered, you can potentially ask one in the Stackoverflow R community.

References

Moeys J (2018) The soil texture wizard: R functions for plotting, classifying, transforming and exploring soil texture data. R package soiltexture Vignette, version 1.5.1.

Venables WN, Smith DM, Team RDC (2009) An introduction to R. Network Theory Limited, Bristol

Chapter 3
Materials

This chapter describes the datasets and R packages used in the book. A total of five datasets are provided and described. They originate from several studies and are made available through a book-associated R package. Most R functions used in this book are either provided in the text or available online in R packages. For some large and complicated functions, we included them in the book-associated R package, which is called `soilspec`. The package contains also the data that are described in the next section. Further instructions to install the package and the set of functions it contains are provided at the end of this chapter.

3.1 Datasets

3.1.1 The Geeves Dataset

The Geeves dataset (Geeves et al. 1994) is a soil survey containing 391 soil samples collected in the wheatbelt of southern New South Wales and northern Victoria, Australia. The area is about 5000 km^2 with a cool temperate climate. The elevation ranges from about 40 to 320 metres above sea level. The main soil types are Chromosols, Dermosols, Kandosols and Sodosols according to the Australian soil classification system or Luvisols, Lixisols and Solonetz according to the WRB system (Minasny et al. 2009). The soil samples were collected between 0 and 100 cm during the year 1994. A map of the spatial locations of the soil samples is provided in the supplementary material in Tang et al. (2020). Each soil sample was air-dried (40°C), crushed and passed through a 2 mm sieve.

The reflectance spectra was obtained in the range 350–2500 nm at 1 nm resolution using an AgriSpec^{TM} instrument with a contact probe (Analytical Spectral Devices, Boulder, Colorado, USA). A Spectralon (Labsphere Inc., North Sutton, N.H., USA) was used as reflectance standard. Each spectrum in the dataset is a

A. M. J.-C. Wadoux et al., *Soil Spectral Inference with R*, Progress in Soil Science,
https://doi.org/10.1007/978-3-030-64896-1_3

wavelength-average of 40 scans. Besides averaging, the spectra used in this book have not been pre-processed.

Soil organic carbon (SOC), clay, silt and sand were analysed in the laboratory using standard methods. SOC was analysed by the dry-combustion method (Nelson and Sommers 1996) and is reported in g 100 g^{-1}. The values of the SOC range from 0.06 to 12.74 g 100 g^{-1}. The soil does not contain inorganic carbon. The hydrometer method (Gee and Bauder 1986) was used for the particle size distribution, and their relative quantity was converted to mass percent. Clay content varies from 5% to 74%, silt content from 2% to 54% and sand content from 14% to 91%.

3.1.2 Soil Mineral Reference Spectra

The soil mineral reference spectra used in this book is a set of 12 spectra of specific mineral compounds measured by diffuse-reflectance laboratory spectrometers in the wavelength range 350–2500 nm. The reference spectra (Kokaly et al. 2017) are made freely available by the USGS through their website. The minerals are purified material so that each reference spectrum contains unique spectro-chemical links related to chemical structure of the mineral. We collected the reference spectra at the Base Spectra library for clay minerals including kaolinite (KGa-2), illite (GDS4), smectite (SWy-1), kaolinite-smectite 50/50 mixture (H89-FR-2), goethite (GDS240) and haematite (GDS576) (Clark et al. 2007). We did not apply any pre-processing to the spectra. The full specifications and methods for scanning are detailed in Kokaly et al. (2017).

3.1.3 Soil Spectra and Colour

The soil spectra and colour dataset is a small set of vis-NIR spectra (wavelength range 350–2500 nm) collected during the Summer of 2013 from a soil profile located at the DEDJTR Rutherglen Centre. The Rutherglen area, situated North-East of Australia in the state of Victoria, is best known for its viticulture industry. The vis-NIR spectra were collected *in situ* with measurements made every 10 cm down to profile. In addition to the spectra, a coloured picture of each soil sample was taken in the laboratory.

3.1.4 Spectra for Standardization

This dataset comprises vis-NIR spectra of a standard material described in Ben Dor et al. (2015). The specific sample is called 'Lucky Bay' and contains (white-coloured) sand material collected from a homogeneous sand dune situated at

Lucky Bay in southwestern Australia. This standard material can be requested from the authors of (Ben Dor et al. 2015) and comes with an associated vis-NIR spectrum measured with their instrumentation and using their published protocol. The instrument and protocol are used to benchmark other spectrometers. The first spectrum of this dataset is the benchmark spectrum. The other two spectra of the dataset were collected with two different vis-NIR spectrometers: (1) Analytical Spectral Devices (ASD) AgricSpec^{TM} spectroradiometer and (2) Spectral Evolution PSR+ 3500 field portable spectroradiometer. Both spectrometers collect high-resolution vis-NIR spectra at 1 nm resolution with a spectral range of 350–2500 nm.

3.1.5 Spectra with Moisture

This dataset contains a subset of the Geeves soil survey data in which the soil samples were scanned at different soil moisture contents. This dataset can be used to illustrate how to remove the moisture effect from infrared spectra using the study of Minasny et al. (2011). A total of 100 soil samples were randomly selected (out of 391) and wetted to reach a sticky point. The samples were then dried in the laboratory for 24 hours. There are three sets of spectra data with different moisture content, categorized as follows:

- absorbance spectra for soil after being wetted (average moisture 12%).
- absorbance spectra for wetted soil after being air-dried for 1 day (average moisture 9%).
- absorbance spectra for soil under air-dried condition (average moisture 5%).

3.2 R Packages

Recall that a package is installed by the `install.packages(" ")` function and loaded by `library()`. Once loaded into the R session, it is possible to obtain information of the packages and help on the functions using:

```
# display the description of the stats package
packageDescription("stats")

# open the documentation and information on the package
help(package = "stats")
```

In addition to `soilspec`, to follow the contents of this book, some other existing packages also need to be installed and loaded. This can be done with the following code.

```
# specify all the packages used in the book
myPackages <- c("asdreader", "RColorBrewer","wavethresh", "MASS",
                "pracma", "plyr", "signal", "SDMTools",
                "tripack", "resemble", "splancs", "sp",
                "scatterplot3d", "prospectr", "RcppArmadillo", "matrixStats",
                "clhs", "viridis", "viridisLite", "caret",
                "ggplot2", "soiltexture","randomForest","pls",
                "Cubist", "lattice", "robustbase", "mvoutlier",
                "devtools")

# define which packages are not installed in the current computer
notInstalled <- myPackages[!(myPackages%in%installed.packages()[ , "Package"])]

# install the missing packages
if(length(notInstalled)>0) install.packages(notInstalled)
```

3.2.1 Soil Science and Pedometrics

aqp The `aqp` package stands for algorithms for quantitative pedology (Beaudette et al. 2013) and was first released in CRAN in 2010. It is the most active pedometrics R package and has had more than 20 updates since its creation. The package has functions to represent the soil profile in three dimensions (spatially and by depth) in a systematic way and functions for numerical classification (e.g. by attribute-space distance), discrimination, clustering and visualization.

The package description and documentation is available online at https://cran.r-project.org/web/packages/aqp/.

soiltexture The `soiltexture` package contains functions for soil texture plotting, categorization and transformation. The package aims at providing tools for converting soil particle size fractions to texture classes, to transform soil texture classes between classification systems around the world and to plot a soil texture triangle.

The package description and documentation is available online at https://cran.r-project.org/web/packages/soiltexture/.

3.2.2 Spectroscopy

ChemoSpec The `ChemoSpec` package was created by Bryan A. Hanson from DePauw University in 2011. `ChemoSpec` brings together a set of functions for analysis of spectral data from different spectrometers such as nuclear magnetic resonance, infrared or Raman. The package provides functions for exploratory analysis, visualization, basic spectra correction, hierarchical clustering and other dimensional reduction tools. One objective of the package is to be user-friendly by providing large sets of examples.

The package description and documentation is available online at https://cran.r-project.org/web/packages/ChemoSpec/.

spectacles The `spectacles` package was previously called `inspectr`. `spectacles` brings together a set of functions for dealing with spectral data (with and emphasis in soil data). This package implements the `Spectra` and `SpectraDataFrame` object classes in an object-oriented programming (OOP; see Bengtsson 2003) framework. The package focuses on the manipulation of spectral data and their associated laboratory measurements and enables the user to apply their own algorithms, pre-processing and filtering as well as exporting visualizations as ggplot objects that can be customized afterwards.

The package description and documentation is available online at https://cran.r-project.org/web/packages/spectacles/.

chemometrics The `chemometrics` package is the companion to the book *Introduction to Multivariate Statistical Analysis in Chemometrics* (Varmuza and Filzmoser 2016). This package includes a selection of functions covering different aspects of spectral data manipulation, including sampling considering the multivariate nature of spectra, outlier detection, filtering, pre-processing and visualization.

The package description and documentation is available online at https://cran.r-project.org/web/packages/chemometrics/.

ChemometricsWithR The `ChemometricsWithR` package was written as a companion of the book *Chemometrics with R – Multivariate Data Analysis in the Natural Sciences and Life Sciences* (Wehrens 2011). The package was removed from CRAN for some time but reappeared in 2019 as the second edition of the book was soon to be released.

The package description and documentation is available online at https://cran.r-project.org/web/packages/ChemometricsWithR/.

prospectr The `prospectr` package is written entirely by soil scientists and provides a large set of functions for spectra correction and pre-processing. The package is also an important reference for the implementation of sampling designs applied to infrared spectra.

The package description and documentation is available online at https://cran.r-project.org/web/packages/prospectr/.

hyperSpec The `hyperSpec` package covers different types of algorithms for hyperspectral, visible, near- and mid-infrared, nuclear magnetic resonance or X-ray fluorescence spectroscopy, among others. This package also offers an OOP framework with S4 classes objects called `hyperSpec`. As with many other packages, the main functionality of the package covers pre-processing, sampling, dimensional reduction, plotting and modelling.

The package description and documentation is available online at https://cran.r-project.org/web/packages/hyperSpec/.

resemble Implementation of a memory-based learning algorithm, where close samples (based on a particular type of distance, usually Mahalanobis) are selected for the modelling operation to create a local model. The package `resemble` has also functions to compute spectral similarity/dissimilarity, e.g. cosine dissimilarity or spectral angle mapper.

The package description and documentation is available online at https://cran.r-project.org/web/packages/resemble/.

simplerspec The package provides a set of functions for spectral data handling and processing using the newly developed R pipe `%>%` operator. The package `simplerspec` contains most common functions for spectra loading and pre-processing, multivariate calibration and plotting. The functions can be combined and integrated in a modelling workflow.

The package description and documentation is available online at https://github.com/philipp-baumann/simplerspec.

mvoutlier The `mvoutlier` package has been developed by the author of the `chemometrics` package. It contains function for multivariate outlier detection based on robust statistics.

The package description and documentation is available online at https://cran.r-project.org/web/packages/mvoutlier/.

3.2.3 Modelling

pls The package `pls` has functions for both principal component analysis and partial least squares (PLS) regression. PLS is probably the most used model for linear dimension reduction.

The package description and documentation is available online at https://cran.r-project.org/web/packages/pls/.

Cubist This package was created by Ross Quinlan as a successor of C4.5 and C5 software packages. The cubist model implemented in `Cubist` is a tree-based model which has each terminal leaves fitted by a linear regression. Cubist has been successfully implemented in spectroscopy for building a model between infrared spectra and associated values of a soil property.

The package description and documentation is available online at https://cran.r-project.org/web/packages/Cubist/.

e1071 The `e1071` package is the first and leading implementation of support vector machines in R. `e1071` also brings together a set of statistical techniques including clustering, multivariate distance calculation and plotting tools.

The package description and documentation is available online at https://cran.r-project.org/web/packages/e1071/.

RandomForest The package provides an implementation of the random forest algorithm, an extension of decision trees. It follows the original implementation from Breiman (2001) and additionally provides functions for computing the model variable importance and to tune the hyperparameters. Note that many implementations of the random forest algorithm are available in R (e.g. `ranger`, `randomForestSRC`, `xgboost` or `Rborist`), each providing different functionalities.

The package description and documentation is available online at https://cran.r-project.org/web/packages/RandomForest/.

signal The `signal` package implements several digital signal processing algorithms originally created in Matlab and its gnu version (Octave). This package is used in spectroscopy for its smoothing filters functions such as the Savitzky-Golay filter.

The package description and documentation is available online at https://cran.r-project.org/web/packages/signal/.

caret The package name is an acronym for classification and regression training. `caret` is a package that implements a unified modelling framework capable of interacting with several other modelling packages. The package activity dates to 2007, and it has been consistently updated over the years. By 2020, this package enables the implementation of more than 230 modelling algorithms.

The package description and documentation is available online at https://cran.r-project.org/web/packages/caret/.

3.2.4 Plotting

base The `base` package is the default plotting set of function in R. The package has a set of plotting functionalities to create high-quality and reproducible two-dimensional visualization with many formatting options.

The package description and documentation is available online at https://cran.r-project.org/web/packages/base/.

ggplot2 The `ggplot2` package is the R implementation of Leland Wilkinson's 'Grammar of Graphics'. This package offers visualization functions based on a structured way of plotting using overlapping layers. It is currently considered as one of the leading plotting packages for data visualization in R. The package description and documentation is available online at https://cran.r-project.org/web/packages/ggplot2/.

3.3 The `soilspec` Book Package

The data and functions used in this book have been compiled into a single R package made available in a GitHub repository. This section explains how to install the package, a brief description of the functions and the datasets it contains.

3.3.1 *Installing the Package*

This R package will have a permanent online public copy on a GitHub repository. The user can download and install this package by using the `install_github()` function from the `devtools` package as follows.

```
# load the required package
library(devtools)

# install the soilspec package from GitHub
devtools::install_github("AlexandreWadoux/soilspec", force = FALSE)
```

This package has a set of documented functions with examples and datasets. As any other package, the user can refer to the documentation by using the `help()` function in the R Console. The datasets used in the book also have a description in their respective help files and are briefly described in the following sections.

3.3.2 *Functions*

- `chBLCext`: function to fit a convex hull to the region of interest. The regions of interest are absorbance or reflectance of some secondary clay minerals or iron oxides from vis-NIR spectra. The function contains three arguments and returns five values. The function argument description is obtained by `?soilspec::chBLCext`.
- `css`: function to determine the optimal number of spectra to be sent to the laboratory for soil analysis. This function works by comparing the probability density function (pdf) of the population to that of the sample set to assess the representativeness of the sample. The two pdfs are compared based on the mean Euclidean distance (msd). The function contains eight arguments and returns three values. The function argument description is obtained by `?soilspec::css`.
- `eval`: function to compute of a set of accuracy measures between observed and predicted continuous values or classes. The user must specify the type of variable, either quantitative or categorical. For both types of variable, a set of accuracy measures is reported. The list of accuracy measures is provided by writing `?soilspec::eval`. The function contains three arguments and returns a number of accuracy measures.

- `myImagePlot`: function to plot an image from a matrix. This simple function takes as single argument a matrix and returns a plot of this matrix. The function details are accessed by writing `?soilspec::myImagePlot`.
- `spectra2colour`: this function converts spectra reflectance into RGB and Munsell colours. The function takes a single argument as input, a matrix or data frame of the spectra and returns the colour for both RGB and Munsell charts. The function returns a data frame where each row is the spectrum followed by the colour. The function description is obtained by `?soilspec::spectra2colour`.

3.3.3 Datasets

- `datEPO`: subset of the Geeves soil survey data in which the soil samples have been scanned at varying moisture contents. The dataset is a list with four data frames. Description of each data frame is provided in the R documentation by writing `?soilspec::datEPO` and in Sect. 3.1.5. This dataset is used in Chap. 10.
- `datsoilspc`: spectra and associated values of laboratory-analysed soil properties from the soil samples described in Geeves et al. (1994). The data provided is a data frame with 5 columns and 391 rows. The first four columns contain values of the clay, silt, sand and total carbon (described in Sect. 3.1.1). The fifth column is a matrix with the infrared spectra. The documentation can be accessed by writing `?soilspec::datsoilspc`. The `datsoilspc` dataset is used in almost all chapters of the book.
- `datStand`: vis-NIR spectra of a standard material as described in Ben Dor et al. (2015). The data are provided as a list with four data frames. The documentation and description of each of them are provided in Sect. 3.1.4 and by writing `?soilspec::datStand`. The `datStand` dataset is used in Chap. 10.
- `mineralRef`: a data frame containing specific mineral compounds measured by laboratory spectrometers. This dataset is used in Chap. 6. The data frame contains 13 columns. The first is the wavelength and the others are 12 spectra from mineral compounds. The full description is provided by writing `?soilspec::mineralRef` or by reading Sect. 3.1.2.
- `rutherglenNIR`: small set of vis-NIR spectra collected in the Rutherglen area. The spectra provided in this dataset are used to derive their mineral composition in Chap. 6. The documentation is provided in `?soilspec::rutherglenNIR`.
- `specSoilCol`: small set of vis-NIR spectra with value of the soil horizon and soil type. In addition to the spectra, coloured pictures of the soil samples were made. They are provided in Chap. 6. This enables the comparison of the colour retrieved from the spectra and that from the picture of the soil samples. The documentation is provided in `?soilspec::specSoilCol`.

References

Beaudette DE, Roudier P, O'Geen AT (2013) Algorithms for quantitative pedology: a toolkit for soil scientists. Comput Geosci 52:258–268

Ben Dor E, Ong C, Lau IC (2015) Reflectance measurements of soils in the laboratory: standards and protocols. Geoderma 245–246:112–124

Bengtsson H (2003) The R. oo package-object-oriented programming with references using standard R code. In: Hornik K, Leisch F, Zeileis A (eds) Proceedings of the 3rd international workshop on distributed statistical computing (DSC 2003), Vienna

Breiman L (2001) Random forests. Mach Learn 45:5–32

Clark RN, Swayze GA, Wise RA, Livo KE, Hoefen TM, Kokaly RF, Sutley SJ (2007) USGS digital spectral library splib06a. US Geological Survey

Gee GW, Bauder JW (1986) Particle-size analysis. In: Methods of soil analysis: part 1 Physical and mineralogical methods, vol 5. American Society of Agronomy, Madison, pp 383–411

Geeves GW, Cresswell HP, Murphy BW, Gessler PI, Chartres CJ, Little IP, Bowman GM (1994) Physical, chemical and morphological properties of soils in the wheat-belt of southern NSW and northern Victoria. NSW Department of Conservation; Land Management/CSIRO Division of Soils Occasional Report, CSIRO

Kokaly RF, Clark RN, Swayze GA, Livo KE, Hoefen TM, Pearson NC, Wise RA, Benzel WM, Lowers HA, Driscoll RL, others (2017) USGS spectral library version 7. US Geological Survey

Minasny B, McBratney AB, Bellon-Maurel V, Roger J-M, Gobrecht A, Ferrand L, Joalland S (2011) Removing the effect of soil moisture from NIR diffuse reflectance spectra for the prediction of soil organic carbon. Geoderma 167:118–124

Minasny B, McBratney AB, Pichon L, Sun W, Short MG (2009) Evaluating near infrared spectroscopy for field prediction of soil properties. Soil Res 47:664–673

Nelson DW, Sommers LE (1996) Total carbon, organic carbon, and organic matter. Methods Soil Anal Part 3 Chem Methods 5:961–1010

Tang Y, Jones E, Minasny B (2020) Evaluating low-cost portable near infrared sensors for rapid analysis of soils from south eastern australia. Geoderma Reg 20:e00240

Varmuza K, Filzmoser P (2016) Introduction to multivariate statistical analysis in chemometrics. CRC Press, Boca Raton

Wehrens R (2011) Chemometrics with R: multivariate data analysis in the natural sciences and life sciences. Springer Science & Business Media, Berlin

Chapter 4
Data Handling of Spectra

This chapter explains with examples how to load and handle spectroscopic data in R. Most spectroscopic data are stored in universal and easily readable formats. Once loaded into R, the user must perform some basic data cleaning, visual examination and filtering of redundant information. Once all these basic operations are performed, the user can save the data in a format that will allow subsequent analyses to be implemented.

The set of packages used in this chapter are installed using the lines below. The book-associated `soilspec` package is also required; see Chap. 3 for information on its installation.

```
#specify all the packages used in the chapter and install them if they are not already
myPackages <- c("asdreader", "RColorBrewer")

# define which packages are not installed in the current computer
notInstalled <- myPackages[!(myPackages %in% installed.packages()[ , "Package"])]

# install the missing packages
if(length(notInstalled)) install.packages(notInstalled)
```

4.1 Importing Data

Spectroscopic data are most often saved in the tabular format, either with the `.csv` or `.txt` extension. The `.csv` extension represents tabular where each value from a cell in a row is separated by a comma. The `.csv` is often preferred over other formats because there is no need to specify how one wants to separate the values when saving the file or how to load the table. The `.txt` extension, conversely, is a standard text document. For spectroscopic data, it requires the user to specify the nature of separation between values. They are most often separated by a comma, a semicolon or a tab character (a space), but any separation can be used.

Let us make an example by loading data from a `.csv` or `.txt` document.

A. M. J.-C. Wadoux et al., *Soil Spectral Inference with R*, Progress in Soil Science,
https://doi.org/10.1007/978-3-030-64896-1_4

```
require(soilspec)

# add the path to the .csv data
path <- system.file("extdata", "soilspec.csv", package = "soilspec")

# read a .csv file
soilspec <- read.csv(path)

# display the first ten column names
names(soilspec)[1:10]
```

```
##  [1] "X"             "clay"          "silt"          "sand"          "Total_Carbon"
##  [6] "X350"          "X351"          "X352"          "X353"          "X354"
```

```
# add the path to the .txt data
path <- system.file("extdata", "soilspec.txt", package = "soilspec")

# read a .txt file, using the tab delimiter
soilspec <- read.delim(path,
                       header = TRUE,
                       sep = "\t")

# display the first ten column names
names(soilspec)[1:10]
```

```
##  [1] "X"             "clay"          "silt"          "sand"          "Total_Carbon"
##  [6] "X350"          "X351"          "X352"          "X353"          "X354"
```

We see that the spectra column names start by `X`. This has been added when saving because the column names cannot be numeric. We also notice that the first column is the row number that was saved with the table. This can easily be removed.

```
# delete the first column
soilspec <- soilspec[,-1]
```

To facilitate further processing and visualization of the spectra in the table, we coerce the spectra into a single data frame that we coerce to a single column.

```
# put the spectra into a single dataframe
spec <- soilspec[grep("X", names(soilspec), value = TRUE)]

# remove the spectra from the current dataframe
soilspec <- soilspec[ , -which(names(soilspec) %in% grep("X", names(soilspec),
                                                         value = TRUE))]

# add the spectra to the dataframe
soilspec$spc <- spec
```

The dataset is now much easier to read.

```
names(soilspec)
```

```
## [1] "clay"          "silt"          "sand"          "Total_Carbon" "spc"
```

We would like to remove the X in front of the column names of the spectra wavelengths. This will make easier plotting in a later step.

```
# take each column name from the spectra dataset
oldNames <- grep("X", names(soilspec$spc), value = TRUE)

# remove the "X" and make a numeric vector
wavelength <- as.numeric(substring(grep("X", names(soilspec$spc), value = T), 2, 20))

# change the name of the columns of the spectra
colnames(soilspec$spc) <- wavelength

# display the first ten column names
colnames(soilspec$spc)[1:10]
```

```
##  [1] "350" "351" "352" "353" "354" "355" "356" "357" "358" "359"
```

4.2 Loading ASD Data

Spectroscopic data may be collected and stored in a specific format. One of the most common is the .asd format for spectra collected using ASD spectrometers. The acronym ASD stands for analytical spectral devices, and data stored in the .asd format can be loaded in R using specific functions. The .asd files contain both the spectra and the metadata. We use the implementation provided in the package asdreader (Roudier 2017). The package provides some files for the example (Fig. 4.1).

```
# load the required package
require(asdreader)

# load the path to some example files
pathExampFiles <- asd_file()
print(pathExampFiles) # print author's path to file
```

```
## [1] "C:/Users/awad2791/Documents/R/win-library/3.6/asdreader/extdata/soil.asd"
```

```
# run the function to read the ".asd" files onto R
spec <- get_spectra(pathExampFiles, type = "reflectance")

# plot the spectrum
plot(as.numeric(spec),
     type = "l",
     ylab = "Reflectance",
     xlab = "Index")
```

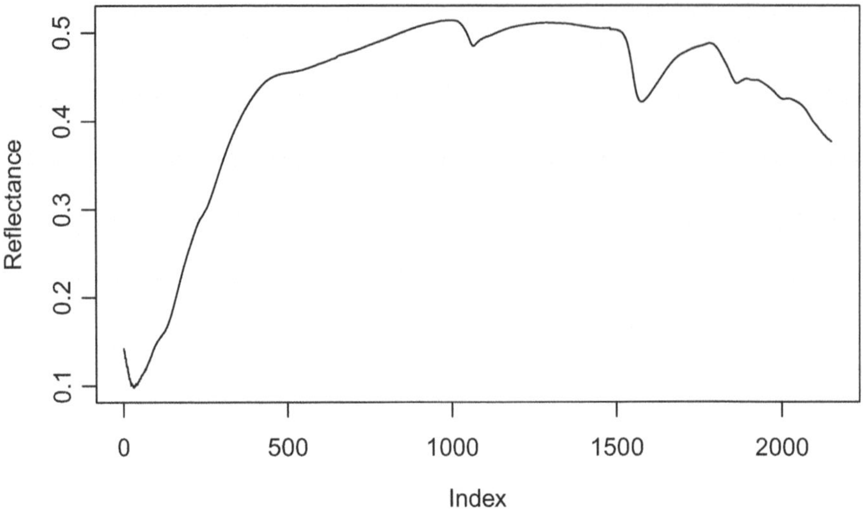

Fig. 4.1 Example spectrum after loading the data with the `asdreader` function. Note that the x-axis is the index and not the wavelength. The information on the wavelength is accessed in the metadata file

We can also obtain the metadata stored along the spectra.

```
# obtain the metadata stored in the ".asd" file
specMeta <- get_metadata(pathExampFiles)

# display the information contained in the metadata
names(specMeta)
```

```
##  [1] "co"                  "comments"          "when"
##  [4] "program_version"     "file_version"      "dc_corr"
##  [7] "dc_time"             "data_type"         "ref_time"
## [10] "ch1_wavel"           "wavel_step"        "data_format"
## [13] "channels"            "it"                "fo"
## [16] "dcc"                 "calibration"       "instrument_num"
## [19] "ip_numbits"          "flags"             "dc_count"
## [22] "ref_count"           "sample_count"      "instrument"
## [25] "bulb"                "swir1_gain"        "swir2_gain"
## [28] "swir1_offset"        "swir2_offset"      "splice1_wavelength"
## [31] "splice2_wavelength"
```

Note that other functions exist for loading `.asd` data into R. Some functions are specific to a spectrometer. This means that the `asdreader` function may not work for all files saved in the `.asd` format and that adaptation of the source codes is sometimes required. As an example, the package `prospectr` provides another function called `readASD` for loading `.asd` data, but only for the spectra acquired with an ASD FieldSpec Pro (ASDI, Boulder, CO) spectroradiometer.

4.3 Plotting the Spectra

We now want to plot the spectra contained in `soilspec$spc`. This can be done using the `graphics` package with the `matplot` function. The `matplot` function plots the columns of one vector (i.e. the wavelength) against the columns of a matrix, in our case the spectral reflectance values.

The data `soilspec$spc` have already been prepared in a file and are saved in the `soilspec` package. We can load it using `data("datsoilspc")` and plot the spectra (Fig. 4.2).

```
# load the example data
data("datsoilspc")

# plot example spectra
matplot(x = colnames(datsoilspc$spc), y = t(datsoilspc$spc),
        xlab = "Wavelength /nm",
        ylab = "Reflectance",
        type = "l",
        lty = 1,
        col = rgb(red = 0.5, green = 0.5, blue = 0.5, alpha = 0.3))
```

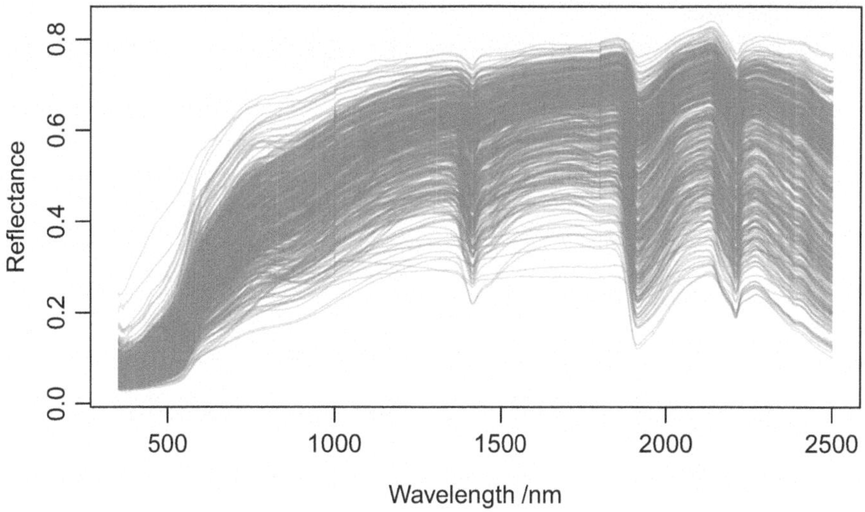

Fig. 4.2 Example set of 391 soil spectra collected by an ASD FieldSpec in the wheatbelt of Australia, covering approximately a 5,000 km^2 area (see also Geeves et al. 1994)

In the matplot function, the arguments `x` and `y` represent the variables to be plotted on the x- and y-axes. The `xlab` and `ylab` arguments specify the names of the x- and y-axes, `type = "l"` stands for line (`type = "p"` for points), `lty = 1` means that we plot solid lines, and `col` is the colour of the lines. In our case, we use the additional `rgb` function to specify the exact colour that we want, given the rgb values and a transparency value (`alpha = 0.3`). We would like to add colours to differentiate spectra with high or low values of the clay content. The clay content is in the column `datsoilspc$clay` and ranges from 5% (low clay content) to 74% (high clay content). We use the colour palette from the `RColorBrewer` package (Fig. 4.3).

```
library(RColorBrewer)

# take three colours from the "RdBu" colour scale
cols = brewer.pal(3, "RdBu")

# make a function which convert the three colours to a gradient of colours
pal = colorRampPalette(cols)

# order the values of clay from the dataframe to find their increasing order
order = findInterval(datsoilspc$clay, sort(datsoilspc$clay))

# plot spectra
matplot(x = colnames(datsoilspc$spc), y = t(datsoilspc$spc),
        xlab = "Wavelength /nm",
        ylab = "Reflectance",
        type = "l",
        lty = 1,
        col = pal(nrow(datsoilspc))[order])

# add a legend to the plot
legend("topleft",
       col = pal(2),
       pch = 19,
       legend = c("Low clay content", "High clay content"))
```

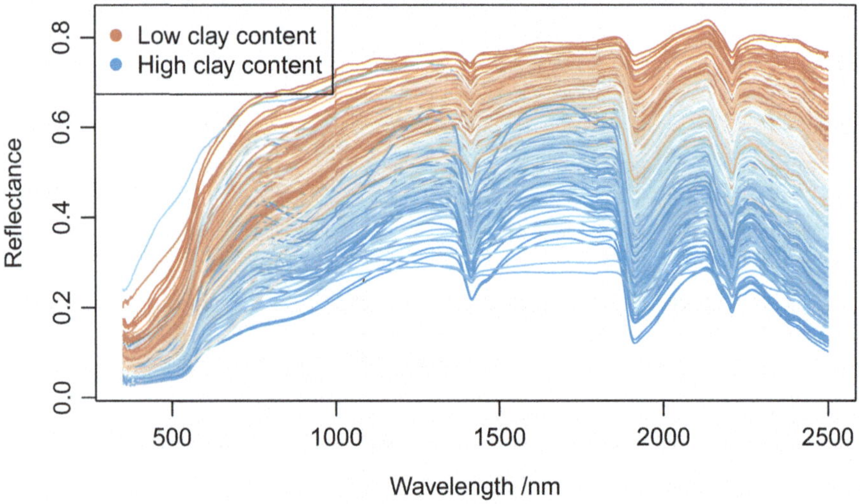

Fig. 4.3 Reflectance soil spectra from Geeves et al. (1994) where the colour represents the amount of clay (low to high clay content) derived from conventional laboratory soil analysis

We can further plot a specific area of the spectra by specifying the argument `ylim`. Let us plot the wavelength range of kaolinite (a clay mineral) at around 2078–2267 nm with red for low clay content and blue for high clay content (Fig. 4.4).

```
#Plot Kaolin bands
matplot(x = colnames(datsoilspc$spc), y = t(datsoilspc$spc),
        xlab = "Wavelength /nm",
        ylab = "Reflectance",
        type = "l",
        lty = 1,
        col = pal(nrow(datsoilspc))[order],
        xlim = c(2078, 2267))
```

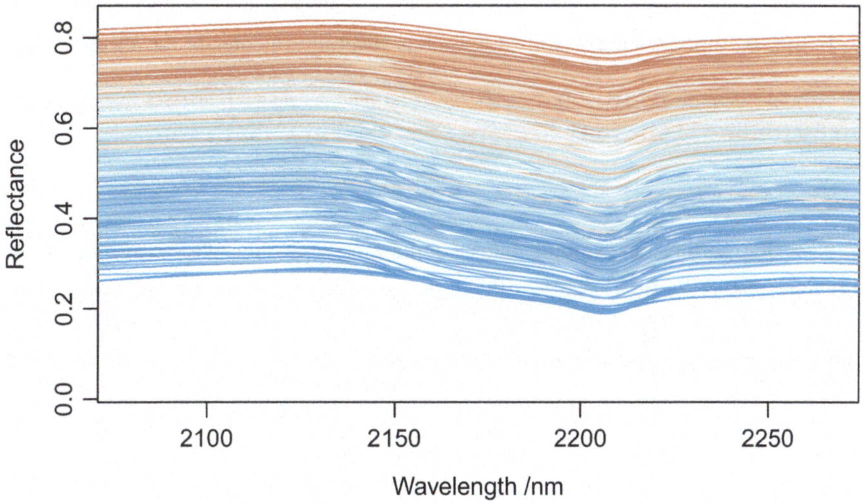

Fig. 4.4 Spectral reflectance range of the kaolinite absorption (2078–2267 nm). The colour represents the amount of clay (low to high clay content) contained in the soil sample associated to the spectra

4.4 Averaging the Replicates

In many cases, each soil sample may be scanned several times (replicates). It is generally good practice to do this to account for soil heterogeneity, natural instrument measurement variations and human error during data collection. In practice, one may want to merge the replicates to smooth out the spectrometer error. Several packages provide functions to aggregate the replicates of the spectra, such as the `spectacles` package. In this example, we will aggregate the spectra using the base `stats` package and the `aggregate` function.

Let us try the function by taking the mean of replicates. Since our dataset does not have replicates, we make an example by assuming that our dataset is composed of spectra collected on 23 different soil samples (23 is convenient because it is a multiple of 391, the total number of spectra available). We therefore start by assigning a number related to each soil sample.

```
# prepare an example, assign 23 different id to our spectral data
# each id is a soil sample
sampId <- rep(1:23, each=nrow(datsoilspc)/23)
datsoilspc$id <- sampId

# in this example, we assume that we have spectra from 23 different soil samples
unique(datsoilspc$id)
```

```
## [1]  1  2  3  4  5  6  7  8  9 10 11 12 13 14 15 16 17 18 19 20 21 22 23
```

In this example, each `id` (soil sample) has several replicates of the spectra.

```
# show number of replicates for the first soil sample
nrow(datsoilspc[datsoilspc$id == 1,])
```

```
## [1] 17
```

Now that we have an example dataset containing replicates, we can average these replicates using the `aggregate` function.

```
# average the replicates of the spectra by soil sample id
datsoilspcAvSpc <- aggregate(datsoilspc$spc,
                             # average using the id as variable
                             by = list(datsoilspc$id),
                             # specify the data
                             data = datsoilspc$spc,
                             # specify the function
                             # we take the mean of the replicates
                             FUN = mean)

# ensure that the number of row is equal the number of soil samples
nrow(datsoilspcAvSpc)
```

```
## [1] 23
```

We can now plot the replicate-averaged spectra (Fig. 4.5).

```
# plot spectra averaged by replicates
matplot(x = colnames(datsoilspcAvSpc[,-1]), y = t(datsoilspcAvSpc[,-1]),
        xlab = "Wavelength /nm",
        ylab = "Reflectance",
        type = "l",
        lty = 1,
        col = rgb(red = 0.5, green = 0.5, blue = 0.5, alpha = 0.3))
```

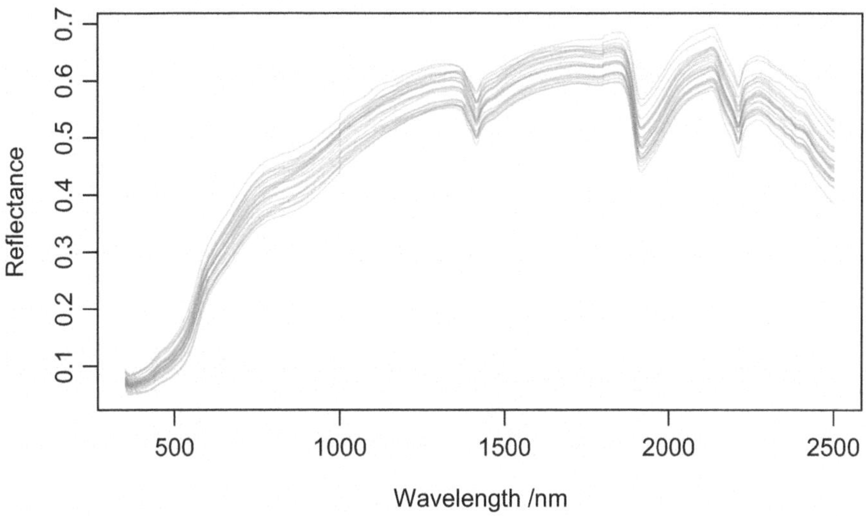

Fig. 4.5 Example set of 23 reflectance spectra (one for each soil sample) obtained after averaging 17 replicates that were available for each soil sample

4.5 Converting Units of Measurement

Spectrometers quantify the amount of energy absorbed by a soil sample. This is measured in terms of light intensity as a function of the wavelength. Spectra are usually provided in terms of the latter. Depending on the type of spectra, one may encounter differ units of measurements. For example, the vis-NIR spectrum is often reported in micrometre (μm) or nanometre (nm). Conversion is made as follows:

- 1 micrometre (μm) = 1000 nanometres (nm)
- 1 nanometre (nm) = 0.001 micrometre (μm)

Let us, for example, convert the aggregated spectra from nm to μm. The values of the column names need to be divided by 1000.

```
# take column names as a numeric vector
namesNm <- as.numeric(colnames(datsoilspcAvSpc[,-1]))

# convert from nanometer to micrometer
namesUm <- namesNm/1000
```

Let us now plot the spectra with the new units of measurement in μm (Fig. 4.6).

```
# plot spectra averaged by replicates
matplot(x = namesUm, y = t(datsoilspcAvSpc[,-1]),
        xlab = expression("Wavelength /" * mu * m),
        ylab = "Reflectance",
        type = "l",
        lty = 1,
        col = rgb(red = 0.5, green = 0.5, blue = 0.5, alpha = 0.3))
```

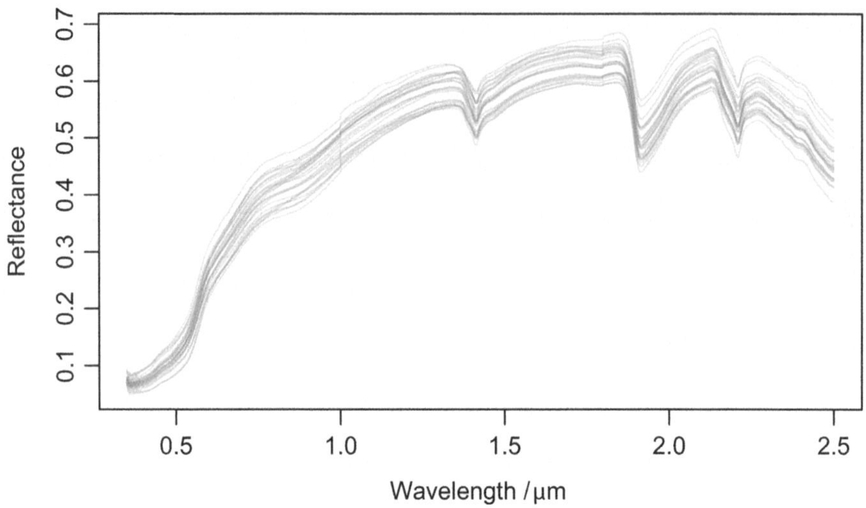

Fig. 4.6 Example set of 23 reflectance spectra with the measurement units in μm

In addition to the wavelength, some spectra are reported in wavenumber. The wavenumber is the reciprocal of the wavelength, i.e. how many wavelengths fit into one unit of distance (i.e. cm). Mid-infrared spectra are usually reported in inverse wavelength (i.e. cm^{-1}). Conversion from wavelength to wavenumber is made by $\lambda = 1/k$ where λ is the wavelength and k is the wavenumber, while conversion from wavenumber to wavelength is made by $k = 1/\lambda$. The wavenumber k has units of distance (e.g. cm^{-1}). What is important to realize is that the conversion does not change the distance measure. If the wavenumber is in cm^{-1}, then the wavelength will be in cm, or if the wavelength is in nm, the wavenumber will be in nm^{-1}. For example, to convert the wavelength 350 nm to cm^{-1}, the nm unit should first be converted to cm: 1 nm = 1^{-7} cm; hence, $350 \times 1^{-7} = 0.000035$. 350 nm is equal to 0.000035 cm which in cm^{-1} is $1/0.000035 = 28571.43$. Let use convert the nm to cm^{-1} from the aggregated spectra.

```
# take column names as a numeric vector
namesNm <- as.numeric(colnames(datsoilspcAvSpc[,-1]))

# convert from nanometre to cm-1
namescm <- namesNm*1e-7
namescmInv <- 1/namescm
```

Let us now plot the spectra with the new units measurement in cm^{-1} (Fig. 4.7).

```
# plot spectra averaged by replicates
matplot(x = namescmInv, y = t(datsoilspcAvSpc[,-1]),
        xlab = expression("Wavenumber /" * cm^-1),
        ylab = "Reflectance",
        type = "l",
        lty = 1,
        col = rgb(red = 0.5, green = 0.5, blue = 0.5, alpha = 0.3))
```

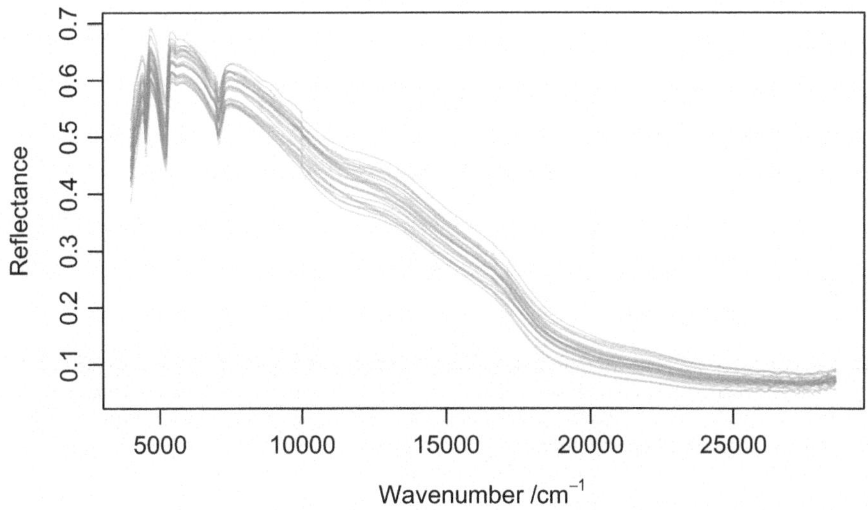

Fig. 4.7 Example set of 23 reflectance spectra with the measurement units in cm^{-1}

4.6 Exporting the Spectra

Once the basic steps of loading the spectra into R and visualizing and averaging the replicates are completed, if necessary, one can also save the spectra to use it at later stage. Several options exist to save the spectra. We present three of them: exporting as a `.txt`, `.csv` or `.Rdata` file. The `.txt` and `.csv` have already been presented earlier in this chapter. We use the same base package to export the spectra (data frame or matrix format).

```
# export the spectra (data) to a .txt file
write.table(datsoilspcAvSpc,
            # path and name of the file (with extension .txt)
            file = "./datsoilspcavspc.txt",
            # tab-separated values
            sep = "\t",
            # decimal separator
            dec = ".",
            # write also the column name
            col.names = TRUE)

# export the spectra (data) to a .csv file
write.csv(datsoilspcAvSpc,
          file = "./datsoilspcavspc.csv")
```

A convenient way to export datasets in R is to use the `.Rdata` extension. Multiple objects can be saved (with the specified `.Rdata` file formats) and will not be altered in any way when loading them back in R. One drawback is that files with `.Rdata` extensions are difficult to open from another software (such as Matlab).

The files are saved using the `save()` function and loaded in R using the `load()` function.

```
# save a file using the .Rdata extension
save(datsoilspcAvSpc,
     file = "./datsoilspcavspc.Rdata")
```

References

Geeves GW, Cresswell HP, Murphy BW, Gessler PI, Chartres CJ, Little IP, Bowman GM (1994) Physical, chemical and morphological properties of soils in the wheat-belt of southern NSW and northern Victoria. NSW Department of Conservation; Land Management/CSIRO Division of Soils Occasional Report, CSIRO

Roudier P (2017) Asdreader: Reading ASD Binary Files in R package version 0.1–3

Chapter 5
Pre-processing of Spectra

Spectral pre-processing techniques aim at improving the quality of the spectra before using them for qualitative analysis or calibration and estimation of soil properties. When scanning the soil, the sample volume, sample preparation, measurement method and measuring parameters such as the choice of scanning time and scanning resolution bring inevitable errors to the recorded spectral data. Further, when scanning a material with non-uniform particle sizes such as soils, the reflectance spectrum is often accompanied by scattering noise.

There are several spectral pre-processing methods available in the literature to ensure that we can use the spectra for inference. The few that are described in this chapter are most often used in soil spectroscopy. A useful reference that describes in more detail each of these pretreatments is Buddenbaum and Steffens (2012).

Spectral pre-processing is the first step of spectral data analysis, and possibly the most important. The reader must be aware that there is no single best method or sequence of methods for spectral pre-processing. The most suitable methods will depend, among others, on the quality of the generated (i.e. raw) spectra and on the sensitivity of the subsequent analysis to random variation in the spectra. A standard analysis is an iterative procedure in which the impact of the spectral pre-processing technique is tested jointly with the subsequent analysis (e.g. output of a multivariate calibration).

In this chapter, we provide examples based on various R packages available for the purpose of signal and spectral pre-processing. The techniques presented range from trimming, simple noise removal and derivative techniques to specific peak removal, baseline correction or continuum removal. This topic has already been extensively covered in the literature, either in textbooks (e.g. Wehrens 2011, Chapter 3) or in R package specific vignettes (e.g. in Stevens and Ramirez-Lopez 2014).

The set of packages used in this chapter are installed using the lines below. The book-associated `soilspec` package is also required; see Chap. 3 for information on its installation.

A. M. J.-C. Wadoux et al., *Soil Spectral Inference with R*, Progress in Soil Science,
https://doi.org/10.1007/978-3-030-64896-1_5

```
# specify all the packages used in the chapter and install them if they are
not already
myPackages <- c("wavethresh", "MASS", "pracma", "plyr",
                "signal", "prospectr", "RcppArmadillo")

# define which packages are not installed in the current computer
notInstalled <- myPackages[!(myPackages %in% installed.packages()[ , "Package"])]

# install the missing packages
if(length(notInstalled)) install.packages(notInstalled)
```

In this chapter, we use the raw visible and near-infrared (vis-NIR) spectra provided by the book-associated `soilspec` package.

```
# load the required package
library(soilspec)

# load the data
data("datsoilspc")
```

Some important details about these spectra are available when executing the script: `?datsoilspc`. Recall that these spectra are vis-NIR spectra of soils collected from the agricultural zone of southern New South Wales and northern Victoria in Australia (Geeves et al. 1994). We get some sense of the data we are working with by plotting them (Fig. 5.1).

```
# plot the spectra using matplot
matplot(x = colnames(datsoilspc$spc), y = t(datsoilspc$spc),
        xlab = "Wavelength /nm",
        ylab = "Reflectance",
        ylim = c(0, 1),
        type = "l",
        lty = 1,
        col = rgb(red = 0.5, green = 0.5, blue = 0.5, alpha = 0.3))
```

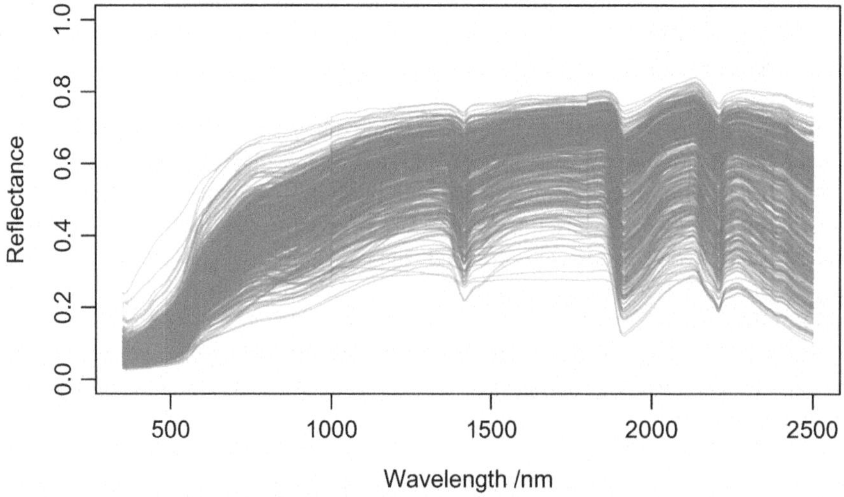

Fig. 5.1 Example set of 391 reflectance vis-NIR soil spectra from the `soilspec` package

Converting data from reflectance values to absorbance values

When the spectra are used to extract soil information by means of modelling, the spectra are better expressed in terms of absorbance rather than reflectance units (Gobrecht et al. 2015) so that the chemical components are linearly related to the spectral wavelengths. Transforming from reflectance to absorbance units is a non-linear calculation that is expressed as either $\log(1/\mathbf{X}^R)$ or $-\log(\mathbf{X}^R)$ (natural logarithms are used here) where $\mathbf{X}^R$ is a matrix of size $n \times b$ containing the measured reflectance, where n is the number of spectra (rows) and b is the number of digital spectral wavelengths (columns). The script below does the conversion for every spectrum $\mathbf{x}_i$ in the dataset. We call the new matrix $\mathbf{X}^A$ (which means the absorbance values of the spectra) as `datsoilspc$spcA` (Fig. 5.2).

```
# change from reflectance to absorbance
datsoilspc$spcA <- log(1/datsoilspc$spc)

# plot first spectra
matplot(x = colnames(datsoilspc$spcA), y = t(datsoilspc$spcA),
        xlab = "Wavelength /nm",
        ylab = "Absorbance",
        ylim = c(0, 4),
        type = "l",
        lty = 1,
        col = rgb(red = 0.5, green = 0.5, blue = 0.5, alpha = 0.3))
```

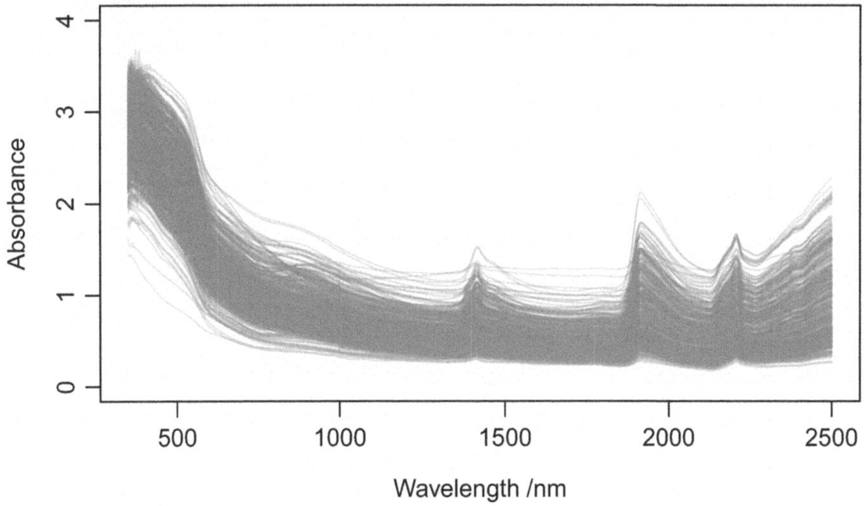

Fig. 5.2 Example set of 391 absorbance soil spectra from the `soilspec` package

5.1 Noise Removal

5.1.1 Spectral Trimming

Trimming spectra is a generic procedure where wavelength ranges with high signal-to-noise ratio are removed. For example, it is a common convention with vis-NIR spectra to remove the wavelengths from 350 to 499 nm (ultraviolet to green) and 2451 to 2500 nm because these do not contain much soil information since they are at the boundary of the range recorded by the sensor. Effectively we want to retain the wavelengths in the spectral range of 500–2450 nm. Without a common procedure for performing such an operation, we need to customize a function or routine ourselves so that this procedure can be easily automated. The function also needs to be generalizable, i.e. the specification of wavelength regions of interest needs to be flexible, and complimentary to the input spectral data.

We plot a single spectrum to see whether we need to remove the noisy spectra at wavelengths from 350 to 499 nm and 2451 to 2500 nm.

```
plot(colnames(datsoilspc$spcA), datsoilspc$spcA[1, ],
     type = "l",
     ylab = "Absorbance", xlab = "Wavelength /nm",
     col = rgb(red = 0.5, green = 0.5, blue = 0.5, alpha = 1))

# identify the area with large noise
rect(xleft = 300, xright = 450,
     ybottom = 2.3, ytop = 2.58,
     border = "red", lwd = 1, lty = "dashed")
```

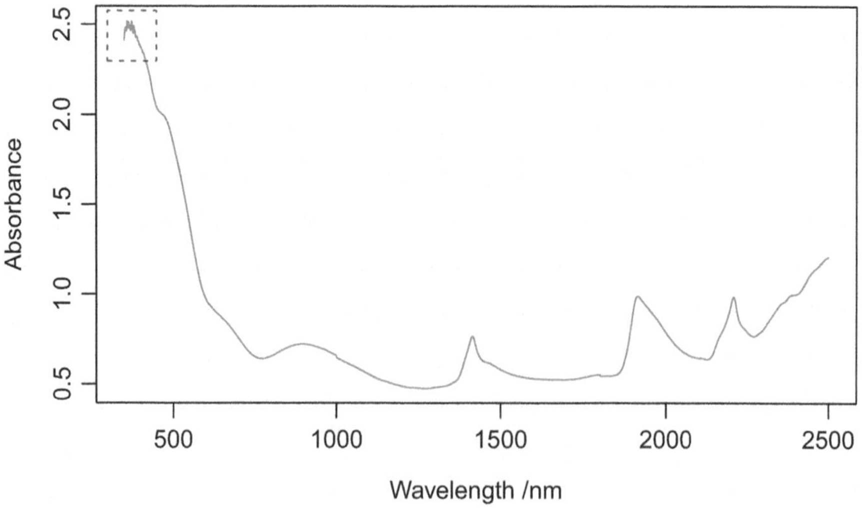

Fig. 5.3 First spectrum of the dataset provided in the `soilspec` package. The red rectangle is an example of area with low signal-to-noise ratio carrying little to no information on soil properties

Figure 5.3 clearly shows some noise at the higher and lower range of this spectrum. In the next step, we build a function to remove these noisy portions.

A function is just a set of general instructions for given specific inputs that when executed will return a given output or outputs. For R beginners, there is no need to try to interpret the function below. Note, however, that it requires two inputs: `spectra` (the dataset upon which we want to apply the spectral trimming) and `wavlimits` (the region of interest that we want to extract). The trimming function is called `trimSpec`.

```
# function for trimming spectra or targeting a specific spectral region of interest
trimSpec <- function(spectra, wavlimits){

  datawavs <- as.numeric(colnames(spectra))
  # set the limits
  limits <- which(datawavs %in% wavlimits)

  # mention the index that we keep from the matrix
  keptIndex <- seq(limits[1], limits[2], 1)

  # keep the index selected previously
  trimmedSpectra <- spectra[, keptIndex]

  # return the trimmed spectra
  keptNames <- datawavs[keptIndex]
  colnames(trimmedSpectra) <- keptNames

  return(trimmedSpectra)
}
```

For our purposes, we want to extract all the spectral data between and including 500 and 2451 nm. Note that we specify this region of interest (`wavlimits`) as `range(500:2451)` which equates simply the integers `500` and `2451`.

```
# trimming the absorbance spectra
datsoilspc$spcAT <- trimSpec(spectra = datsoilspc$spcA, wavlimits = range(500:2451))
```

Plotting the spectrum as before is the same, except that now we need to specify the new wavelength sequence of the trimmed spectra (Fig. 5.4).

```
# create a new wavelength sequence
wavs <- seq(500, 2451, by = 1)

# plot the trimmed spectra
plot(wavs, datsoilspc$spcAT[1, ], type = "l",
     ylab = "Absorbance", xlab = "Wavelength /nm",
     col = rgb(red = 0.5, green = 0.5, blue = 0.5, alpha = 1))
```

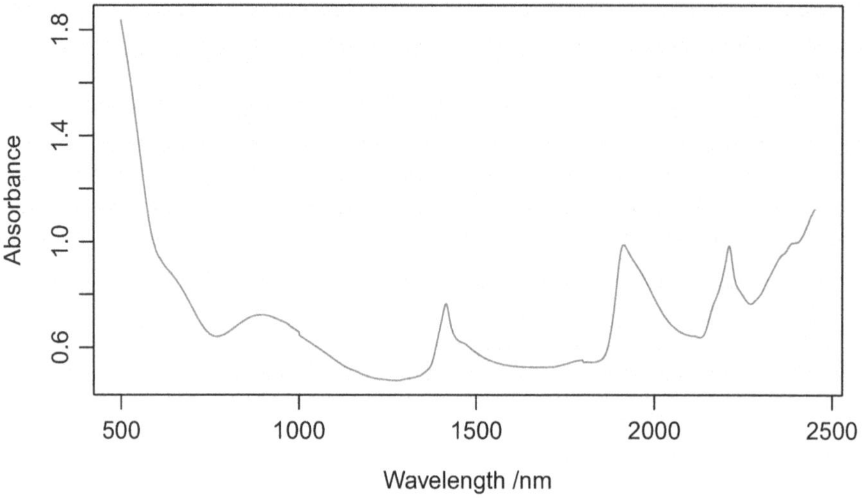

Fig. 5.4 First absorbance spectrum of the dataset provided in the `soilspec` package. This spectrum is obtained after removing the wavelengths smaller than 500 nm and greater than 2451 nm

5.1.2 *Moving Window Average*

In a moving window average operation, each wavelength value is taken as the average of the neighbouring wavelengths. The original spectra are smoothed, which reduces the information content but also the noise that it contains too. The user must specify the size of the window, i.e. over how many wavelengths the values are averaged. The larger the window size, the more important the smoothing of the whole spectrum. Note that the wavelengths at the beginning and end of the spectra will be lost in the process. The number of wavelengths lost is obtained by $(w-1)/2$ where w is the window size.

To illustrate the smoothing effect, we add random noise to the trimmed spectrum derived in the previous section.

```
# add some random noise to simulate an example of noisy spectra
datsoilspc$spcAtNoisy <- datsoilspc$spcAT + rnorm(ncol(datsoilspc$spcAT), 0, 0.003)
```

The package `prospectr` offers a fast computation of the moving average.

```
# load the required package
library(prospectr)

# specify the window size
windowMa <- 11

# apply the moving average
datsoilspc$spcAtMa <- movav(X = datsoilspc$spcAtNoisy, w = windowMa)
```

We can now plot the smoothed spectrum obtained using a moving window of size 11. For illustration, we colour the original noisy spectra as red (Fig. 5.5).

```
# plot the noisy spectrum
plot(names(datsoilspc$spcAtNoisy[1,]), datsoilspc$spcAtNoisy[1,],
     type = "l",
     ylab = "Absorbance", xlab = "Wavelength /nm",
     col = rgb(red = 1, green = 0, blue = 0, alpha = 1))

# add the smoothed spectrum
lines(names(datsoilspc$spcAtMa[1,]), datsoilspc$spcAtMa[1, ],
      col = rgb(red = 0.3, green = 0.3, blue = 0.3, alpha = 1))

# add a legend
legend("topright",
       legend = c("raw", "moving average"),
       lty = c(1, 1),
       col = 2:1)
```

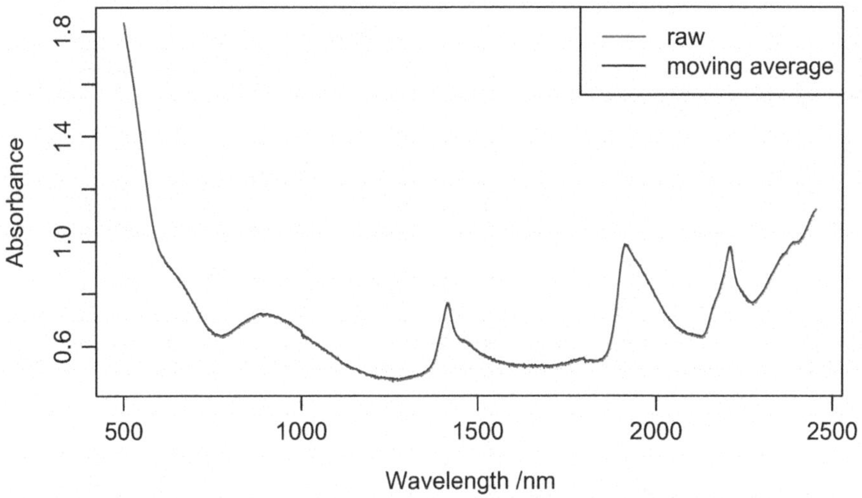

Fig. 5.5 Noisy (red line) and smoothed (black line) first absorbance spectrum of the `datsoilspec` dataset. The smoothed spectrum is obtained by a moving window average of size 11

We can zoom in to highlight the effect of the smoothing (Fig. 5.6).

```
# plot the noisy spectrum
plot(names(datsoilspc$spcAtNoisy[1,]), datsoilspc$spcAtNoisy[1,],
     type = "l",
     ylab = "Absorbance", xlab = "Wavelength /nm",
     col = rgb(red = 1, green = 0, blue = 0, alpha = 1),
```

```
     xlim = c(1250,1750), ylim = c(0.4,0.8))

# add the smoothed spectrum
lines(names(datsoilspc$spcAtMa[1,]), datsoilspc$spcAtMa[1, ],
      col = rgb(red = 0.3, green = 0.3, blue = 0.3, alpha = 1))

# add a legend
legend("topright",
       legend = c("raw", "moving average"),
       lty = c(1, 1),
       col = 2:1)
```

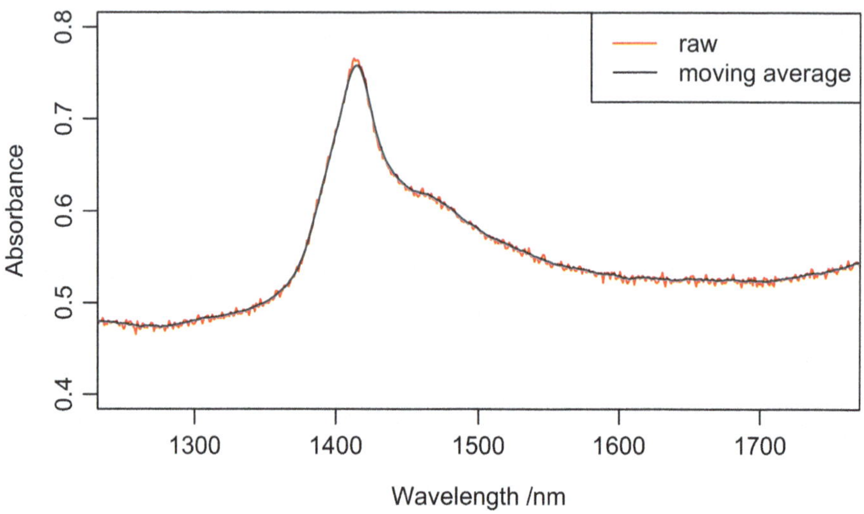

Fig. 5.6 Noisy (red line) and smoothed (black line) first absorbance spectrum in the range 1250–1750 nm of the `datsoilspec` dataset. The smoothed spectrum is obtained by a moving window average of size 11

5.1.3 Savitzky-Golay Filtering

The Savitzky-Golay filter (Savitzky and Golay 1964) fits a local polynomial regression (of order *k*) on a series of spectral values to determine the smoothed value for each wavelength, for a given filter length (also called the window size).

In the following example, we use a second-order polynomial and a filter length of 11. Note that the type of polynomial and window size can be changed to suit user specifications. The general R scripting form of the Savitzky-Golay filter is shown below. It is one of the most widely used pre-processing techniques, and it is therefore implemented in most packages on spectroscopic analysis, e.g. the `savitzkyGolay` function in `prospectr` (Stevens and Ramirez-Lopez 2014).

```
# function for applying Savitzky-Golay smoothing filter
filterSg <- function(spectra, w, k, m) {
  spectra <- as.matrix(spectra)

  ## run filter, the window size in the sgolayfilt function is called n and
  ## the polynomial order is called p
  sg <- aaply(spectra, 1, sgolayfilt, n = w, p = k, m = m)

  ## arrange appropriately if a single sample
  if (nrow(spectra) == 1) {
    sg <- matrix(sg, dim(spectra))
  }

  ## return data frame
  sg <- as.data.frame(sg)
  colnames(sg) <- colnames(spectra)

  return(sg)
}
```

Essentially this function uses the `sgolayfilt` function from the `signal` package. It uses the `aaply` function from the `plyr` package to apply the filter over the entire spectra collection. The parameters include the spectra; a value for w, i.e. the window size; a value for k, i.e. the polynomial order; and a value for m which is whether the derivative of the polynomial is returned. To return the first derivative, we would set the value of m to 1. However, we just want to filter our spectra without returning the derivatives, in which case we set m to 0. We can now run the function.

```
# load R libraries necessary to use filterSg function
library(signal)
library(plyr)

# filter spectra dataset
datsoilspc$spcAtSg <- filterSg(datsoilspc$spcAtNoisy,
                               w = 11,
                               k = 2,
                               m = 0)
```

We can plot the smoothed spectrum on top of the original noisy trimmed spectra (Fig. 5.7).

```
# plot the noisy spectrum
plot(names(datsoilspc$spcAtNoisy[1,]), datsoilspc$spcAtNoisy[1,],
     type = "l",
     ylab = "Absorbance", xlab = "Wavelength /nm",
     col = rgb(red = 1, green = 0, blue = 0, alpha = 1))

# add the smoothed spectrum
lines(names(datsoilspc$spcAtSg[1,]), datsoilspc$spcAtSg[1, ],
      col = rgb(red = 0.3, green = 0.3, blue = 0.3, alpha = 1))

# add a legend
legend("topright",
       legend = c("raw", "Savitzky-Golay"),
       lty = c(1, 1),
       col = 2:1)
```

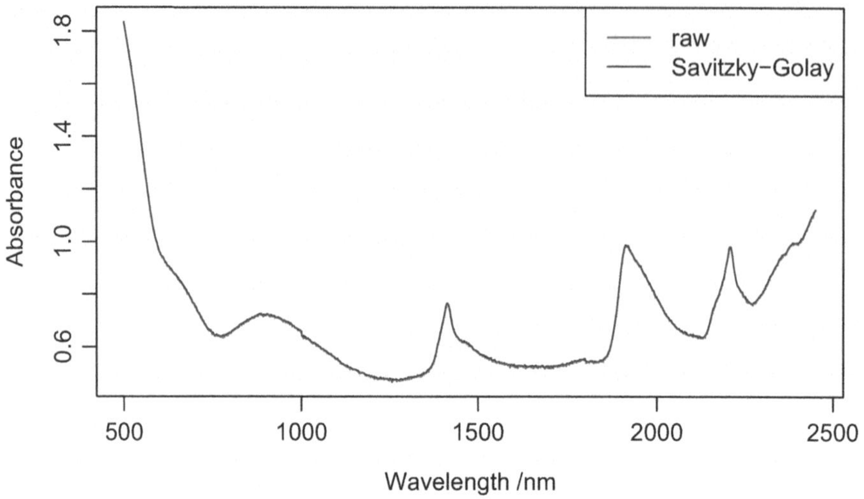

Fig. 5.7 Noisy (red line) and smoothed (black line) first absorbance spectrum of the `datsoilspec` dataset. The smoothed spectrum is obtained by a Savitzky-Golay smoothing filter with a window size of 11 and a polynomial of order 2

5.2 Scatter Correction

Light scattered by soil samples results in deviations dependent on the wavelength, path length and sensitivity of the detector (Siesler et al. 2008). The deviations are difficult to quantify during soil scanning. Most often, a pre-processing step is used to correct for deviations due to light scattering. The simplest method is to centre each individual spectrum to zero and to then divide each spectral band value by the standard deviation of the whole spectrum. It is assumed that the standard deviation correlates with deviations due to light scattering (e.g. path length). A different approach is multiplicative light scattering (MSC), where each spectrum is shifted and scaled to an ideal reference spectrum. A reference spectrum is, for example, a 'correct' scan of the sample not affected by light scattering. In practice, this reference spectrum is unknown, and we use the mean spectrum of the spectra as a reference. Finally, a trend can be removed from each spectrum, which leaves out the unexplained variation of the model for the analysis. These three techniques are described in the next section with their implementation in R.

5.2.1 Standard Normal Variate

Standard normal variate transform (SNV) corrects for single light scattering (Barnes et al. 1989). SNV which is also known as z-transformation or as centring and scaling (operating per spectrum or row-wise) normalizes each spectrum (which we defined

$\mathbf{x}_i$) to zero mean and unit variance by subtracting the mean of the spectrum ($\mu_{\mathbf{x}_i}$) and dividing the difference by its standard deviation ($\sigma_{\mathbf{x}_i}$):

$$\mathbf{x}_i^{snv} = \frac{\mathbf{x}_i - \mu_{\mathbf{x}_i}}{\sigma_{\mathbf{x}_i}}, \tag{5.1}$$

where $\mathbf{x}_i^{snv}$ is the SNV corrected spectrum. This equation can be applied to each soil spectrum one at a time. Merged into a function, it can be scripted as follows:

```
# function for applying standard normal variate transformation
snvBLC <- function(spectra) {
  spectra <- as.matrix(spectra)
  snvMat <- matrix(NA, ncol = ncol(spectra), nrow = nrow(spectra))

  # apply the standardization to each row
  for (i in 1:nrow(spectra)) {
    snvMat[i, ] <- (spectra[i, ] - mean(spectra[i, ]))/sd(spectra[i,])
  }
  snvMat<- as.data.frame(snvMat)
  colnames(snvMat) <- colnames(spectra)

  return(snvMat)
}
```

This function performs SNV for all the spectra in the collection and outputs a new SNV transformed spectra dataset. All it requires is the spectra table for input.

```
# apply the standard normal variate transformation
datsoilspc$specSnvC <- snvBLC(datsoilspc$spcAT)
```

We can plot the scatter-corrected spectra (Fig. 5.8).

```
# plot the scatter-corrected spectra
plot(names(datsoilspc$specSnvC[1,]), datsoilspc$specSnvC[1,],
     type = "l",
     ylab = " ", xlab = "Wavelength /nm",
     col = rgb(red = 1, green = 0, blue = 0, alpha = 1))

# plot the original spectra
lines(names(datsoilspc$spcAT[1,]), datsoilspc$spcAT[1, ],
      col = rgb(red = 0.3, green = 0.3, blue = 0.3, alpha = 1))

# add a legend
legend("topright",
       legend = c("raw", "SNV"),
       lty = c(1, 1), col = 1:2)
```

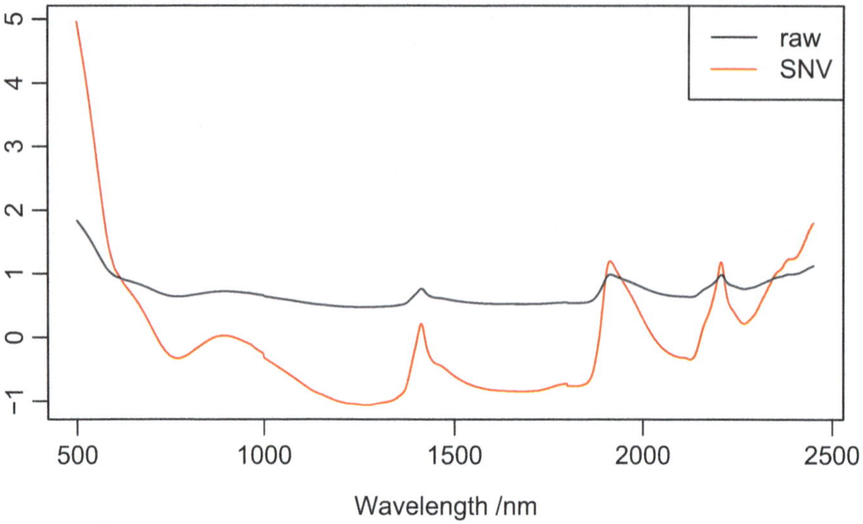

Fig. 5.8 Scatter-corrected (red line) and original (black line) first absorbance spectrum of the `datsoilspec` dataset

Note that Fearn (2008) recommends applying SNV scatter correction after filtering the noise from the spectra (see previous section).

5.2.2 Multiplicative Scatter Correction

Multiplicative scatter correction (MSC) is used to compensate for multiplicative deviations dependent from the wavelength. The correction aligns each spectrum to a reference spectrum so that baseline and amplification effects are at the same average level in every spectrum (Isaksson and Næs 1988; Naes et al. 1990). As this reference spectrum is unknown, the mean spectrum of a given spectral library, denoted $\mathbf{x}_{\text{ref}}$ hereafter, is used. This spectrum represents the mean scattering and offset (Helland et al. 1995). Each spectrum $\mathbf{x}_i$ is then fitted to the reference spectrum using the least squares method:

$$\mathbf{x}_i = a_i + b_i \mathbf{x}_{\text{ref}} + \boldsymbol{\varepsilon}_i, \tag{5.2}$$

where a and b are the intercept and slope of the linear model and $\boldsymbol{\varepsilon}$ are the residuals for a spectrum i. Ideally $\boldsymbol{\varepsilon}_i$ contains the important information, because scattering and offset are represented by the coefficients a and b. The MSC spectrum is calculated by determining the coefficients for each spectrum and then performing the transformation as follows:

$$\mathbf{x}_i^{msc} = \frac{\mathbf{x}_i - a_i}{b_i}, \tag{5.3}$$

where $\mathbf{x}_i^{msc}$ is the spectrum corrected for multiplicative scatter. The MSC function can be scripted as follows:

```
# function for applying multiplicative scatter correction
mscBLC <- function(spectra) {

  # first calculate a mean spectrum.
  meanSpec <- as.matrix(colMeans(spectra))
  mscMat <- matrix(NA, ncol = ncol(spectra), nrow = nrow(spectra))
  spectra <- as.matrix(spectra)

  # make a loop over each row
  for (i in 1:nrow(spectra)) {

    # determine the slope and intercept coefficients
    specLM <- lm(spectra[i, ] ~ meanSpec)
    specCE <- t(as.matrix(specLM$coefficients))

    # adjust the spectra
    mscMat[i, ] <- t(as.matrix((spectra[i, ] - specCE[1, 1])/specCE[1,2]))
  }
  mscMat <- as.data.frame(mscMat)
  colnames(mscMat) <- colnames(spectra)

  return(mscMat)
}
```

As you may note, the first part of this function generates the mean spectrum `meanSpec`. Then for each spectrum, we derive the coefficients using the `lm` function from the `stats` package. The coefficients are saved in the `specCE` object, which are then used for the correction calculation.

```
# apply the multiplicative scatter correction
datsoilspc$specMscC <- mscBLC(datsoilspc$spcAT)
```

We can plot the scatter-corrected spectra (Fig. 5.9).

```
# plot the multiplicative scatter-corrected spectra
plot(names(datsoilspc$specMscC[1,]), datsoilspc$specMscC[1,],
     type = "l",
     ylab = " ", xlab = "Wavelength /nm",
     col = rgb(red = 1, green = 0, blue = 0, alpha = 1))

# add the original spectra to the plot
lines(names(datsoilspc$spcAT[1,]), datsoilspc$spcAT[1, ],
      col = rgb(red = 0.3, green = 0.3, blue = 0.3, alpha = 1))

# add a legend
legend("topright",
       legend = c("MSC", "raw"),
       lty = c(1, 1), col = 2:1)
```

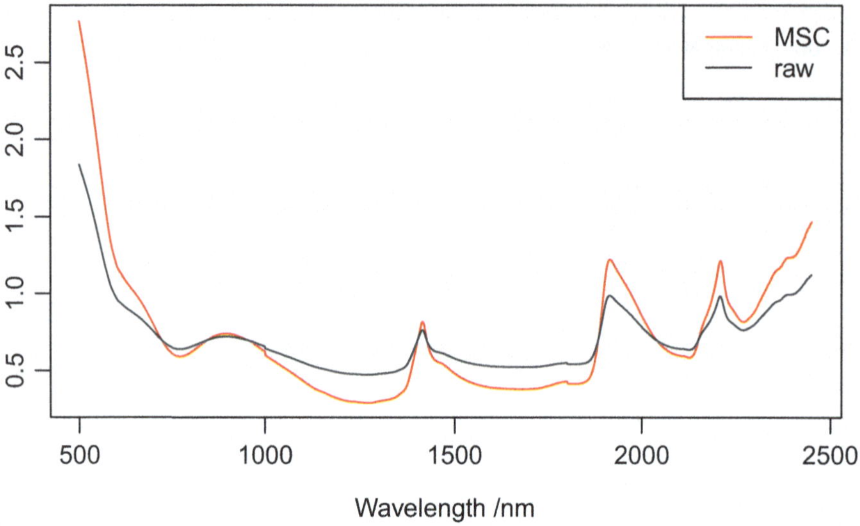

Fig. 5.9 Multiplicative scatter-corrected (red line) and original (black line) first absorbance spectrum of the `datsoilspec` dataset

In general, both SNV and MSC lead to similar results. The advantage of SNV is that there is no requirement to provide a reference spectrum. MSC, in contrast, is less prone to the spectra noise (Rinnan et al. 2009). A detailed comparison of SNV and MSC is provided by Dhanoa et al. (1994).

5.2.3 Detrending

A practical alternative to SNV and MSC is to remove the mean value or a linear trend from the spectra. This method is called detrending. This method can be used in combination with SNV or MSC. Let us make a function and apply it to each spectrum separately in the matrix. The workhorse function inside our own `detrendSpc` function is the `detrend` function which comes from the `pracma` package.

```
# function for detrending a matrix of spectra
detrendSpc <- function(spectra) {
  # load the required package
  require(pracma)

  detrendMat <- matrix(NA, ncol = ncol(spectra), nrow = nrow(spectra))
  spectra <- as.matrix(spectra)
```

```
  # make a loop over each row
  for (i in 1:nrow(spectra)) {

    # detrend each spectra, specify the linear model
    specLM <- pracma::detrend(spectra[i, ], tt = "linear")

    # take the values and store in the matrix
    detrendMat[i, ] <- as.numeric(specLM[,1])
  }
  detrendMat<- as.data.frame(detrendMat)
  colnames(detrendMat) <- colnames(spectra)

  return(detrendMat)
}
```

We can now apply it to each individual spectrum. Note that in the above function we specify `pracma::detrend` to force R to use the `detrend` function from the `pracma` package. The `detrend` function is also implemented in the `propectr` package but in a different way. In `prospectr`, detrending involves applying a SNV transform and fitting a second-order polynomial model to return the residuals of this model (Mark 1989). In this section, we provide a simpler example.

We can apply the function `detrendSpc` to our matrix as follows (Fig. 5.10).

```
# detrend the spectra
datsoilspc$specDT <- detrendSpc(datsoilspc$spcAT)
```

```
# plot the detrended spectra
plot(names(datsoilspc$specDT[1,]), datsoilspc$specDT[1,],
     type = "l",
     ylab = " ", xlab = "Wavelength /nm",
     col = rgb(red = 1, green = 0, blue = 0, alpha = 1),
     ylim=c(-0.3, 2))

# add the original spectra
lines(names(datsoilspc$spcAT[1,]), datsoilspc$spcAT[1, ],
      col = rgb(red = 0.3, green = 0.3, blue = 0.3, alpha = 1))

# add the remove linear trend
lines(names(datsoilspc$spcAT[1,]),
      datsoilspc$spcAT[1, ] - datsoilspc$specDT[1,],
      lty = 5)

# add a legend
legend("topright",
       legend = c("raw", "detrended", "linear trend"),
       lty = c(1, 1, 5), col = c(1, 2, 1))
```

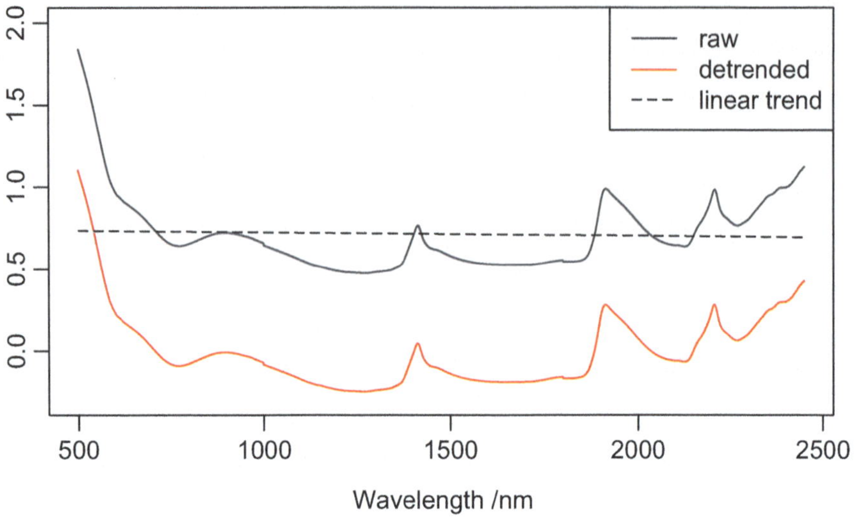

Fig. 5.10 Detrend (red line) and original (black line) first absorbance spectrum of the `datsoilspec` dataset

Performing a detrending before a SNV transform is not recommended (Barnes et al. 1989), but as was mentioned previously, it is possible to apply SNV first and detrend after. In this case, it is better to use MSC because both transformations are applied simultaneously.

5.3 Derivatives

Converting the spectra to first- or second-order derivatives aims at accentuating the absorbance features contained in the spectra. Derivatives also remove both additive and multiplicative effects on the spectra (Rinnan et al. 2009). Taking the first-order derivative detrends the spectrum (see previous section), while taking the second-order derivative both detrends and removes a linear trend (as in the MSC example). Computing the derivative of a spectrum is usually performed after a trimming or initial smoothing. When computing the first- and second-order derivatives with the Savitzky-Golay filter, an initial smoothing is not necessary.

Stevens and Ramirez-Lopez (2014) provide two methods for taking the derivatives. The first is by taking the finite difference between two consecutive wavelength values. This is achieved with the `base` package in R. This method has the major drawback that it tends to increase noise in the spectra. The second method builds on the Savitzky-Golay filtering, by taking the derivative of the smoothing functions to provide means of accentuating the regions of absorbance. The latter method is preferred; we provide a simple implementation below. Alternatively, a third method exists based on the Norris-Williams derivation. This method yields similar derivation to the popular Savitzky-Golay filtering. For parsimony, we will not implement the Norris-Williams method in this chapter.

5.3.1 First- and Second-Order Derivatives

The first- and second-order derivatives of the smoothed spectra are computed using the `filterSg` function, by specifying this time $m = 1$ or $m = 2$ to indicate whether we want to take the first- or second-order derivative of the smoothed spectra, respectively. The window size w also needs to be specified. We use the trimmed spectra obtained from the previous section.

```
# take first derivative of spectra
datsoilspc$specDeriv1 <- filterSg(datsoilspc$spcAT,
                                   w = 11,
                                   k = 2,
                                   m = 1)

# take second derivative of spectra
datsoilspc$specDeriv2 <- filterSg(datsoilspc$spcAT,
                                   w = 11,
                                   k = 2,
                                   m = 2)
```

We can make a plot of the two derivatives of the smoothed spectra (Fig. 5.11).

```
# plot the first order derivative spectra
plot(names(datsoilspc$specDeriv1[1,]), datsoilspc$specDeriv1[1,],
     type = "l",
     ylab = " ",
     xlab = "Wavelength /nm",
     col = rgb(red = 1, green = 0, blue = 0, alpha = 1))

# plot the second order derivative spectra
lines(names(datsoilspc$specDeriv2[1,]), datsoilspc$specDeriv2[1, ],
      col = rgb(red = 0.3, green = 0.3, blue = 0.3, alpha = 1))

# add a legend
legend("topleft",
       legend = c("First order derivative", "Second order derivative"),
       lty = c(1, 1), col = 2:1)
```

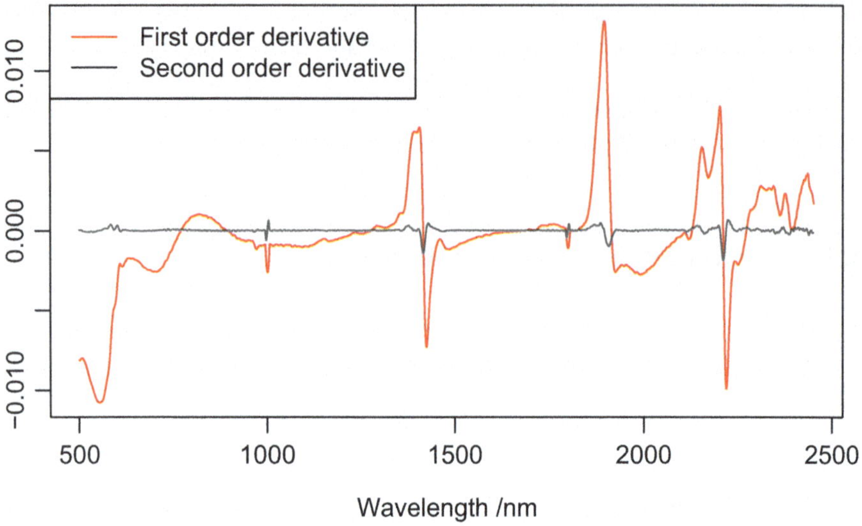

Fig. 5.11 First (red line)- and second (black line)-order derivative of one absorbance spectrum from the `datsoilspc` dataset. The derivative spectra are obtained by the `filterSg` function which uses the Savitzky-Golay filtering

5.4 Centring and Standardizing

Centring and standardizing transforms the spectral values in each wavelength to zero mean or zero mean and unit variance, respectively. Unlike the SNV transformation which standardizes each individual spectrum, centring and standardizing is applied for each wavelength of the spectra (column-wise). Centring of a wavelength is obtained by subtracting the spectral wavelength value by the mean of all the spectra values for this wavelength. Standardization is obtained by subtracting each spectral wavelength value by the mean of all the spectra values for this wavelength and dividing by their standard deviation. It can be simply implemented in the `scale` function from the `base` package.

```
# centre the spectra wavelengths
datsoilspc$specNorm <- scale(datsoilspc$spcAT, center = TRUE, scale = FALSE)

# standardize the spectra wavelengths
datsoilspc$specSdt <- scale(datsoilspc$spcAT, center = TRUE, scale = TRUE)
```

We can now plot with their original spectra (Fig. 5.12).

```
# plot the original spectra
plot(colnames(datsoilspc$spcAT), datsoilspc$spcAT[1,],
     type = "l",
     ylab = " ",
     xlab = "Wavelength /nm",
     col = rgb(red = 1, green = 0, blue = 0, alpha = 1),
```

```
     ylim= c(-1.5, 2.2))

# add to the plot the centred spectra
lines(names(datsoilspc$specNorm[1,]), datsoilspc$specNorm[1,],
      col = rgb(red = 0.3, green = 1, blue = 0.3, alpha = 1))

# add to the plot the standardized spectra
lines(names(datsoilspc$specSdt[1,]), datsoilspc$specSdt[1,],
      col = rgb(red = 0.3, green = 0.3, blue = 1, alpha = 1))

# add a legend
legend("topright",
       legend = c("raw", "centred", "standardized"),
       lty = c(1, 1, 1),
       col = 2:4)
```

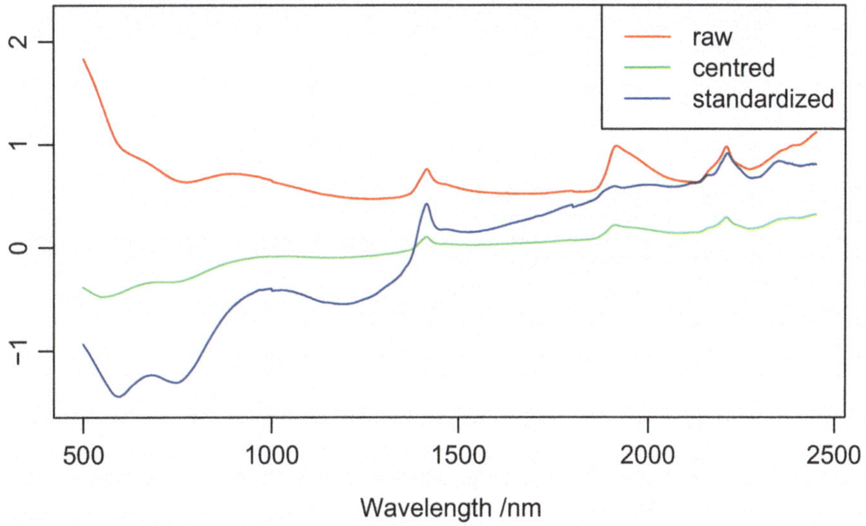

Fig. 5.12 Centred (green line), standardized (blue line) and original (red line) first absorbance spectrum of the `datsoilspec` dataset

5.5 Spectral or Dimension Reduction

5.5.1 Resampling

Despite all the filtering and baseline corrections we may perform, some models we try to use for calibrating soil data with spectra are difficult to fit and validate when each individual spectrum has many wavelengths to consider. We may want to reduce the dimensions of our spectra while keeping the more important absorbance features. One such technique that could be used in tandem with baseline corrections is simple aggregation through resampling and averaging. Essentially with a moving window of given size or wavelength, we calculate the average. The larger the window, the more detail is lost from the spectrum. This approach differs only slightly from the averaging approach described previously in that here the window movement across a spectrum is non-overlapping. Ultimately this means wavebands are removed after calculation of the average, thus reducing the actual dimensions of the spectra.

We provide the following function.

```
# reduce the column dimension of the spectra by averaging
compSpec <- function(spectra, w) {

  # ensure that the number of columns can be divided by the window size
  if(ncol(spectra)%%w != 0)
  {stop("Error: choose a compatible window size")
  }else{
    compMat <- matrix(NA, ncol = (ncol(spectra))/w, nrow = nrow(spectra))
    cc <- 1

    # loop over each column
    for (i in 1:ncol(compMat)) {
      compMat[, i] <- rowMeans(spectra[, cc:(cc + (w - 1))])
      cc <- cc + w
    }

    # provide the new column name (wavelength)
    colab = seq(as.numeric(colnames(spectra)[1]),
                as.numeric(colnames(spectra)[ncol(spectra)]), by = w)
    compMat <- as.data.frame(compMat)
    colnames(compMat) <- colab
  }

  return(compMat)
}
```

The resampling and averaging function is called `compSpec` and is also implemented in the `spectracus` package. It requires two inputs: the spectra that we want to resample and average and the window size w. The window size can be arbitrarily selected. However, if the spectra range (though more importantly the number of columns) is not divisible by the window size, an error will be produced. In this case, you need either to find an appropriate window size or to implement spectral trimming to modify the number of wavelengths. In the following example, we use the `datsoilspc$spcAT` spectra dataset (number of columns $b = 1952$) and use a window size of 8 to reduce the column-wise dimension of the spectra library to 244.

```
# apply the resampling
datsoilspc$specComp <- compSpec(spectra = datsoilspc$spcAT, w = 8)

# dimension of spectra library
dim(datsoilspc$spcAT); dim(datsoilspc$specComp)
```

```
## [1]  391 1952
```

```
## [1] 391 244
```

Note that resampling is also implemented in prospectr. In prospectr, the resample function uses an interpolation (spline or linear regression) to resample the original spectra to new coordinates. This function is particularly suited to match the resolution from one instrument to another because the user must specify the exact resolution to which the spectra have to be resampled.

We can try to resample the trimmed spectra using the resample function from prospectr.

```
# define the current resolution
oldWavs <- as.numeric(colnames(datsoilspc$spcAT))

# define the new resolution, resample every 8 wavelengths
newWavs <- seq(from = min(oldWavs), to = max(oldWavs), by = 8)

# apply the resampling
specResam <- prospectr::resample(X = datsoilspc$spcAT,
                                wav = oldWavs,
                                new.wav = newWavs,
                                interpol = "linear")

# dimension of spectra library
dim(datsoilspc$spcAT); dim(datsoilspc$specComp)
```

```
## [1]  391 1952
```

```
## [1] 391 244
```

We can plot and compare the resampling techniques from our function or that from the prospectr package (Fig. 5.13).

```
# plot the original spectra
plot(colnames(datsoilspc$spcAT), datsoilspc$spcAT[1,],
     type = "l",
     ylab = "Absorbance", xlab = "Wavelength /nm",
     col = rgb(red = 1, green = 0, blue = 0, alpha = 1))

# add to the plot the resampled spectra by our function
lines(names(datsoilspc$specComp[1,]), datsoilspc$specComp[1,],
      col = rgb(red = 0.3, green = 1, blue = 0.3, alpha = 1))

# add to the plot the resampled spectra from prospectr
lines(names(datsoilspc$specResam[1,]), datsoilspc$specResam[1,],
      col = rgb(red = 0.3, green = 0.3, blue = 1, alpha = 1))

# add a legend
legend("topright",
       legend = c("raw", "spectracus resampled", "prospectr resampled"),
       lty = c(1, 1, 1), col = 2:4)
```

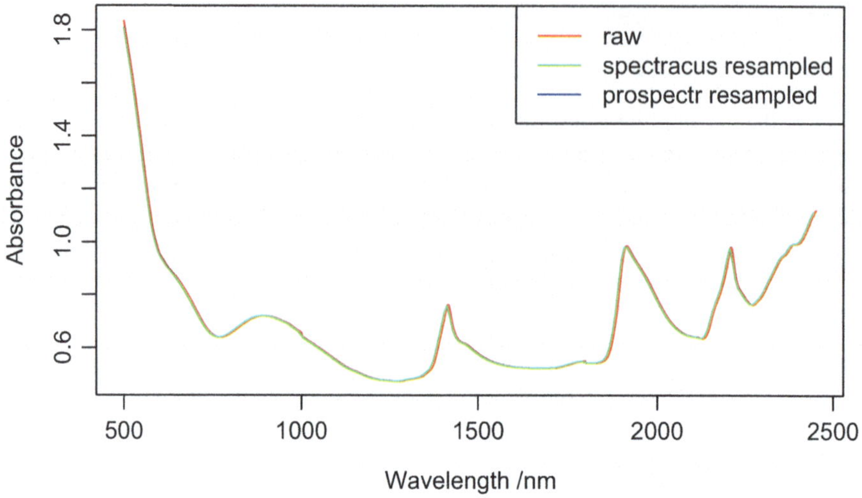

Fig. 5.13 Original (red line), resampled with the `compSpec` function (green line) and resampled using the `resample` function from `prospectr` package (blue line) first absorbance spectrum of the `datsoilspec` dataset

The two methods yield very similar resampled spectra, but a few minor differences are observed.

5.5.2 Wavelets

Wavelets are a naturally useful tool for processing spectra in terms of filtering and dimension reduction. More importantly, with wavelets, we can investigate distinct scales of variation within each spectrum by changing the support of the wavelet filtering function. Some further background on wavelets can be found in Lark and Webster (1999). In the example below, we reduce our spectra down to a few important elements without losing the important absorbance features. We can run the wavelet function over the spectra for which it will compute the wavelet coefficients and a smoothed approximation of the data at each scale. Scale is represented as the support of the filter, where, for example, the increasing dilation of the support will result in smoother and more general representations of the original spectra (Viscarra-Rossel and Lark 2009). However, if we want to, we could reconstruct the original spectra from any scale by multiplying the wavelet coefficients and the smoother signal at each scale. The wavelet function is used to reduce the dimensions of the original spectra down to a few components, i.e. to return the smoothed or more generalized spectra. Here we use the `wavethresh` package and its `wd` or wavelet function. The help file for the wd function describes in more detail the parameters necessary for the wavelet filtering.

We have made a customized function that runs on top of the wd function so that it returns spectra at the requested scale. To run the wavelet implementation, we need to ensure our spectra in terms of numbers of wavebands (the columns) are dyadic. In other words, we just need a number for x in 2^x. In the case of vis-NIR spectra instrumentation where we might expect output to 1 nm spectral resolution data within the 350–2500 nm range (2151 wavelengths), it is most appropriate to find the nearest dyadic number which is 2048 or 2(11). A viable spectral range with which we may work could be between and including 404 and 2451 nm with this particular vis-NIR data. Essentially, we need to do this procedure because the support of the wavelets is dilated by a factor of 2^x for each increasing scale.

Let us first define the function.

```
# function for the wavelet transform of the spectra
filterWl <- function(spectra, res){
  nm2 <- 2^c(1:100)
  vs <- ncol(spectra)
  if (sum(nm2 == vs) != 1) {
    stop("Error: Number of columns in spectra table needs to be power of two")
  }
  waveSpectra <- matrix(NA, ncol = 2^res, nrow = nrow(spectra))
  for (i in 1:nrow(spectra)){
    wds <- wd(as.matrix(spectra[i, ]),
              bc = "symmetric",
              filter.number = 10,
              family = "DaubExPhase",
              min.scale = 2)
    waveSpectra[i, ] <- accessC.wd(wds, level = res)
  }
  wavs <- as.numeric(colnames(spectra))
  colnames(waveSpectra) <- seq((wavs[1] + 0.5 * (length(wavs)/(2^res))),
                               rev(wavs)[1],
                               by = length(wavs)/(2^res))

  return(waveSpectra)
}
```

With the appropriately trimmed spectra, the `filterWl` function requires two inputs: the spectra and an integer (less than 11 in this case) for `res` which indicates the scale that we want to output the smoothed spectra. The example below uses the original unfiltered absorbance spectra `datsoilspc$spcA` from the beginning of this chapter. We will set the resolution `res` to 8. Keep in mind that setting `res` to 11 in this case will potentially return the original spectra table, yet the function does not allow this to occur because it is a redundant operation.

```
require(wavethresh)

# first trimming the datsoilspc$spcA object to 404-2451 spectral range
datsoilspc$spcAT2 <- trimSpec(datsoilspc$spcA, wavlimits = range(404:2451))

# the resolution of the decomposed spectra
res <- 8
# if you want to return the original data specWavelet
# res = 11
```

```
# run wavelet smooth function
datsoilspc$spcAT2Wl <- filterWl(datsoilspc$spcAT2, res = 8)

# dimension of the wavelet transform dataset
dim(datsoilspc$spcAT2Wl)
```

```
## [1] 391 256
```

We can now plot the original and wavelet transformed spectra (Fig. 5.14).

```
# plot the original spectra
plot(names(datsoilspc$spcAT2[1,]), datsoilspc$spcAT2[1,],
     type = "l",
     ylab = "Wavelet values", xlab = "Wavelength /nm",
     col = rgb(red = 1, green = 0, blue = 0, alpha = 1),
     ylim = c(0.2,7))

# add the wavelet transform spectra
lines(names(datsoilspc$spcAT2Wl[1,]), datsoilspc$spcAT2Wl[1, ],
      col = rgb(red = 0.3, green = 0.3, blue = 0.3, alpha = 1))

# add a legend
legend("topright",
       legend = c("Wavelet transformed", "original"),
       lty = c(1, 1),
       col = 2:1)
```

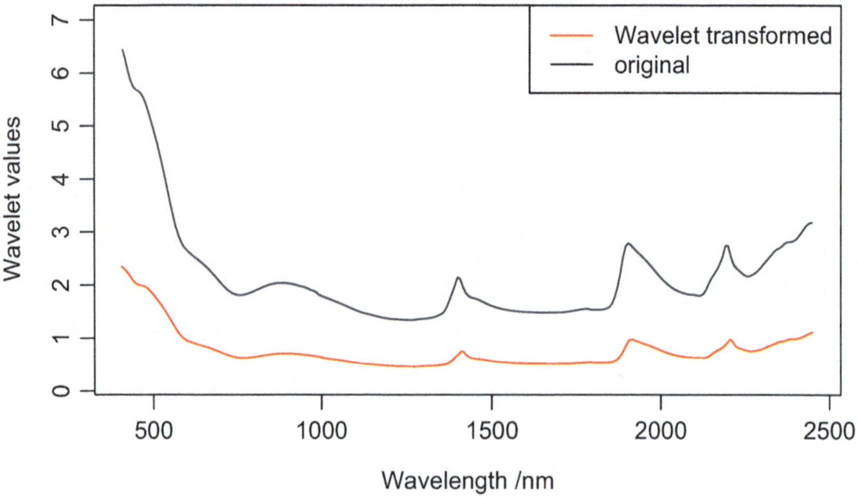

Fig. 5.14 Original (black line) and wavelet transformed (red line) first absorbance spectra of the `datsoilspec` dataset

5.6 Other Specific Transformations

5.6.1 Splice Correction

Some spectra come with a measurement error visible on the spectra. This is common in ASD spectrometers, for example, where there is a transition between different detectors. These transitions commonly occur at 1000 and 1800 nm. Correcting this measurement error is not straightforward because we do not have information about it (otherwise we would have corrected it in the device). This measurement error is often a shift of a specific spectral range. Let us illustrate this (Fig. 5.15).

```
# plot the ten first spectra of the dataset for illustration
matplot(colnames(datsoilspc$spcAT), t(datsoilspc$spcAT[c(1:10),]),
        type = "l",
        ylab = "Absorbance", xlab = "Wavelength /nm",
        col = rgb(red = 0.3, green = 0.3, blue = 0.3, alpha = 0.8))

# identify the area with a systematic shift
rect(xleft = 950, xright = 1050,
     ybottom = 0.2, ytop = 1.2,
     border ="red", lwd=1, lty="dashed")

# identify the area with a systematic shift
rect(xleft = 1800, xright = 1900,
     ybottom = 0.2, ytop = 1.2,
     border ="red", lwd=1, lty="dashed")
```

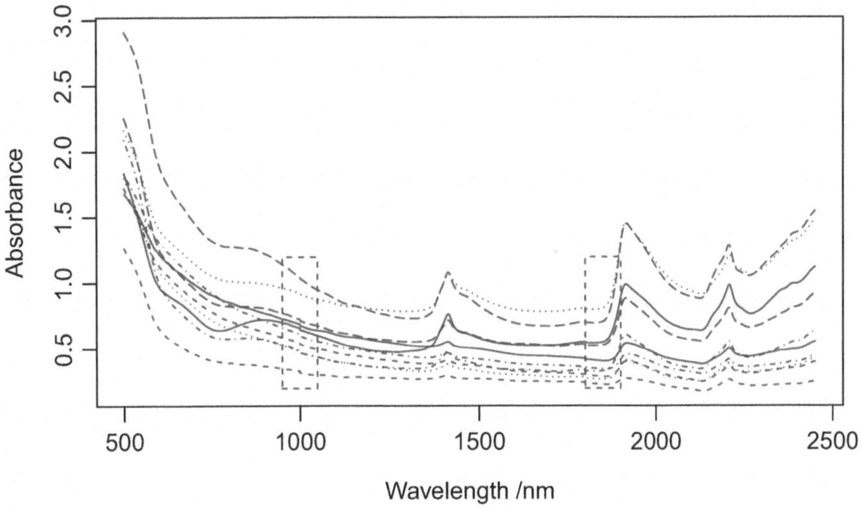

Fig. 5.15 The first ten absorbance spectra of the `datsoilspc` data from the `soilspec` package. The red rectangles identify the spectral range with known systematic shift due to the spectrometer

These shifts are localized between the wavelengths at 1000 and 1800 nm (Fig. 5.16).

```
# plot the spectra around 1000nm
matplot(colnames(datsoilspc$spcAT), t(datsoilspc$spcAT[c(1:10),]),
        type = "l",
        ylab = "Absorbance", xlab = "Wavelength /nm",
        col = rgb(red = 0.3, green = 0.3, blue = 0.3, alpha = 0.8),
        xlim = c(950,1050),
        ylim = c(0.2,1.2))
```

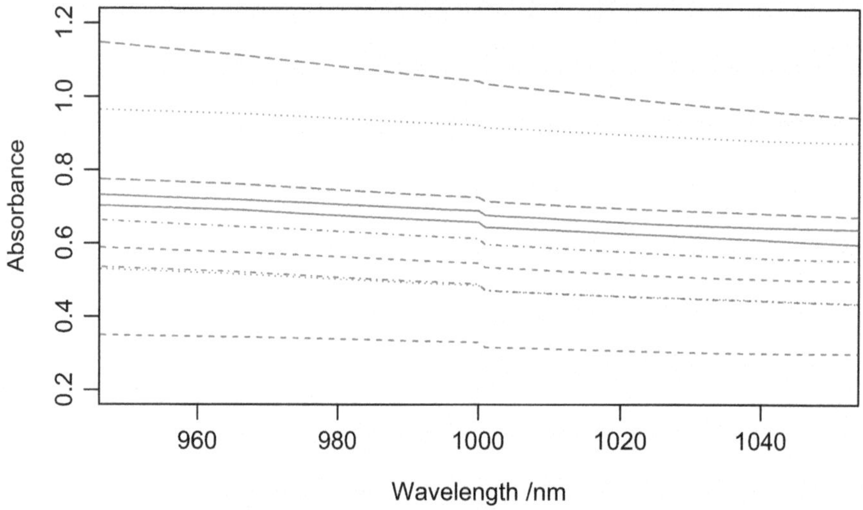

Fig. 5.16 The first ten absorbance spectra of the `datsoilspc` data from the `soilspec` package in the range 950–1050 nm. At around 1000 nm exists a known systematic shift due to the spectrometer

We can use a splice correction implemented in the `prospectr` package with the function `spliceCorrection`. This function performs a linear interpolation from the values located at the edges of the specified shifts.

```
# indicate the exact wavelengths at which the shits are located
sshifts <- c(1000, 1800)

# correct the spectral 'shifts' using the spliceCorrection function
datsoilspc$spcAtSplc <- spliceCorrection(X = datsoilspc$spcAT,
                                         wav = wavs,
                                         splice = sshifts)
```

We can plot the corrected spectra (Fig. 5.17).

```
matplot(x = colnames(datsoilspc$spcAT), y = t(datsoilspc$spcAtSplc),
        xlab = "Wavelength /nm",
        ylab = "Absorbance",
        type = "l",
        lty = 1,
        col = rgb(red = 0.5, green = 0.5, blue = 0.5, alpha = 0.3))
```

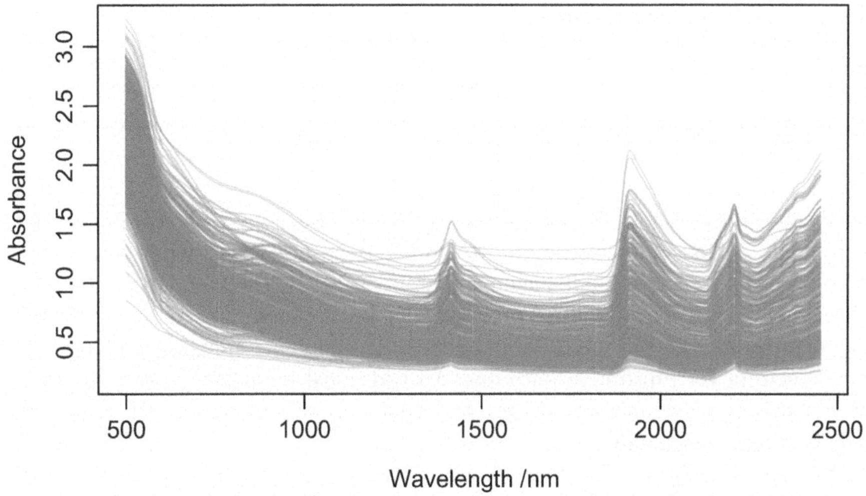

Fig. 5.17 Splice corrected absorbance spectra of the `datsoilspc` dataset from the `soilspec` package

We now plot the corrected spectra around 1000 nm to show the difference with previous plot (Fig. 5.18).

```
# plot the spectra around 1000nm
matplot(colnames(datsoilspc$spcAT), t(datsoilspc$spcAtSplc[c(1:10),]),
        type = "l",
        ylab = "Absorbance", xlab = "Wavelength /nm",
        col = rgb(red = 0.3, green = 0.3, blue = 0.3, alpha = 0.8),
        xlim = c(950,1050), ylim = c(0.2,1.2))
```

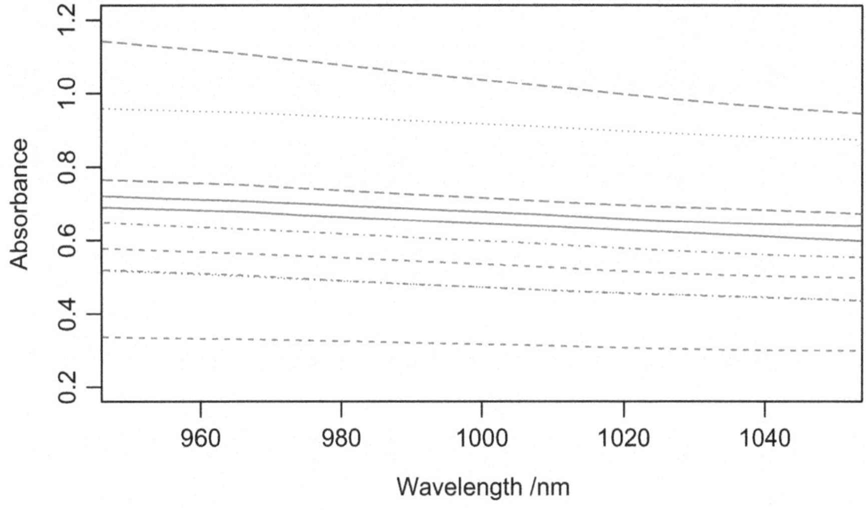

Fig. 5.18 Splice corrected absorbance spectra at the range 950–1050 nm of the `datsoilspc` dataset from the `soilspec` package

5.6.2 Continuum Removal

Convex hull or continuum removal (CR) is one type of baseline method that works by fitting a convex hull to each spectrum and computing the deviations from the hull (Clark and Roush 1984). For absorbance spectra, CR gives value of 0 to all parts of the spectrum that lie on the convex hull and values between 0 and 1 to regions inside absorption bands. The opposite is true if we are dealing with reflectance data. Essentially CR accentuates the absorption bands in the spectra while minimizing brightness differences (Buddenbaum and Steffens 2012). The function for applying CR involves a number of internal processes that are not important for understanding the main points of this algorithm, so we just apply the function and plot the first spectrum for illustrative purposes. An additional parameter also needs to be specified, which we indicate as either `A` if the data is in absorbance units or `R` if the data is in reflectance units.

Note that the convex hull does not always work as intended if applied to the full spectral range of the data because of an issue with defining the hull points and then approximating the associated continuum line. The convex hull is usually applied when we are dealing with discrete regions of a spectrum. In the example below, we will use a function upon a specific NIR wavelength region where the secondary clay mineral kaolinite can be detected. This region is between 2079 and 2277 nm (Clark et al. 2007). We will apply this function to the reflectance data. First we trim the spectral library to the specified region and then apply the `continuumRemoval` function (Fig. 5.19).

```
# first trim reflectance spectra to region
datsoilspc$spcAtR <- trimSpec(spectra = datsoilspc$spc, wavlimits = range(2079:2277))

# apply CR function
datsoilspc$specChcSub <- continuumRemoval(X = datsoilspc$spcAtR, type = "R")

# plot first spectrum
plot(seq(from = 2079,to = 2277,by = 1), datsoilspc$specChcSub[1, ],
     type = "l",
     ylab = "CR units", xlab = "Wavelength /nm",
     lty = 1,
     col = rgb(red = 0, green = 0, blue = 0, alpha = 0.8))
```

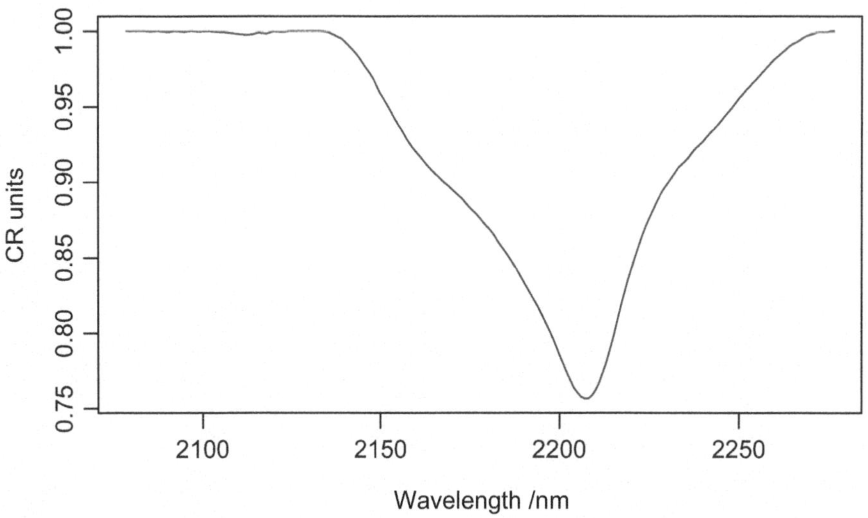

Fig. 5.19 Example of a continuum removed spectra for the range 2079–2277 nm. The continuum removed spectra is obtained with the `continuumRemoval` function from the `prospectr` package

We can also compute it for the whole spectra, but be mindful not to try and interpret anything meaningful from the output as this is just done for illustrative purposes (Fig. 5.20).

```
# apply CR function to the trimmed absorbance spectra (hence type = "A")
datsoilspc$specChC <- continuumRemoval(X = datsoilspc$spcAT, type = "A")

# plot original spectra
plot(wavs, datsoilspc$spcAT[1, ],
     type = "l",
     ylab="Absorbance", xlab= "Wavelength /nm",
     lty = 1,
     col = rgb(red = 1, green = 0, blue = 0, alpha = 1),
     ylim=c(0, 2))

# add to the plot hte continuum removed spectra
lines(wavs, datsoilspc$specChC[1, ],
      col = rgb(red = 0.3, green = 0.3, blue = 0.3, alpha = 1))

# add a legend
legend("topright",
       legend = c("raw", "continuum removal"),
       lty = c(1, 1),
       col = 2:1)
```

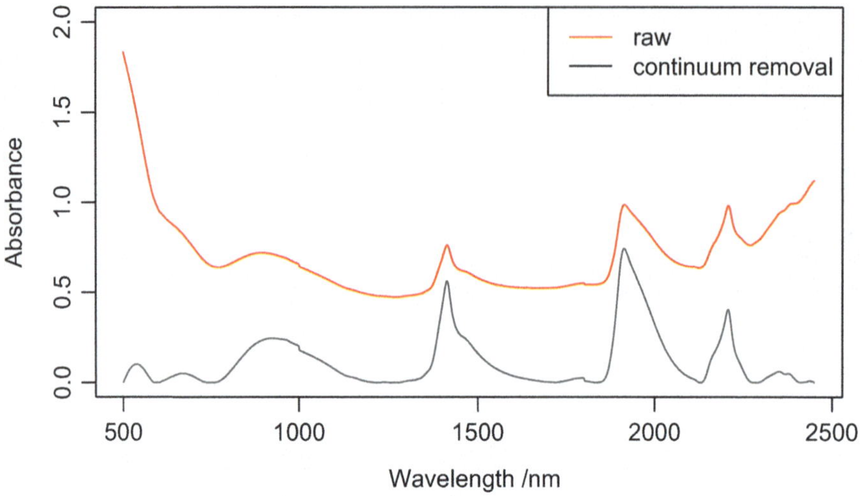

Fig. 5.20 Original (red line) and continuum removed (black line) absorbance spectra from the `datsoilspc` dataset from the `soilspec` package

References

Barnes RJ, Dhanoa MS, Lister SJ (1989) Standard normal variate transformation and de-trending of near-infrared diffuse reflectance spectra. Appl Spectrosc 43:772–777

Buddenbaum H, Steffens M (2012) The effects of spectral pretreatments on chemometric analyses of soil profiles using laboratory imaging spectroscopy. Appl Environ Soil Sci 2012:274903

Clark RN, Roush TL (1984) Reflectance spectroscopy: quantitative analysis techniques for remote sensing applications. J Geophys Res Solid Earth 89:6329–6340

Clark RN, Swayze GA, Wise RA, Livo KE, Hoefen TM, Kokaly RF, Sutley SJ (2007) USGS digital spectral library splib06a. US Geological Survey

Dhanoa MS, Lister SJ, Sanderson R, Barnes RJ (1994) The link between multiplicative scatter correction (MSC) and standard normal variate (SNV) transformations of NIR spectra. J Near Infrared Spectrosc 2:43–47

Fearn T (2008) The interaction between standard normal variate and derivatives. NIR News 19:16–17

Geeves GW, Cresswell HP, Murphy BW, Gessler PI, Chartres CJ, Little IP, Bowman GM (1994) Physical, chemical and morphological properties of soils in the wheat-belt of southern NSW and northern Victoria. NSW Department of Conservation; Land Management/CSIRO Division of Soils Occasional Report, CSIRO

Gobrecht A, Bendoula R, Roger J-M, Bellon-Maurel V (2015) Combining linear polarization spectroscopy and the representative layer theory to measure the Beer-Lambert law absorbance of highly scattering materials. Anal Chim Acta 853:486–494

Helland IS, Næs T, Isaksson T (1995) Related versions of the multiplicative scatter correction method for preprocessing spectroscopic data. Chemom Intell Lab Syst 29:233–241

Isaksson T, Næs T (1988) The effect of multiplicative scatter correction (MSC) and linearity improvement in NIR spectroscopy. Appl Spectrosc 42:1273–1284

Lark RM, Webster R (1999) Analysis and elucidation of soil variation using wavelets. Eur J Soil Sci 50:185–206

Mark H (1989) Chemometrics in near-infrared spectroscopy. Anal Chim Acta 223:75–93
Naes T, Isaksson T, Kowalski B (1990) Locally weighted regression and scatter correction for near-infrared reflectance data. Anal Chem 62:664–673
Rinnan Å, Van Den Berg F, Engelsen SB (2009) Review of the most common pre-processing techniques for near-infrared spectra. TrAC Trends Anal Chem 28:1201–1222
Savitzky A, Golay MJE (1964) Smoothing and differentiation of data by simplified least squares procedures. Anal Chem 36:1627–1639
Siesler HW, Ozaki Y, Kawata S, Heise HM (2008) Near-infrared spectroscopy: principles, instruments, applications. Wiley, Weinheim
Stevens A, Ramirez-Lopez L (2014) An introduction to the prospectr package
Viscarra-Rossel RA, Lark RM (2009) Improved analysis and modelling of soil diffuse reflectance spectra using wavelets. Eur J Soil Sci 60:453–464
Wehrens R (2011) Chemometrics with R: multivariate data analysis in the natural sciences and life sciences. Springer Science & Business Media, Berlin

Chapter 6
Exploratory Soil Spectral Analysis

The previous chapters focused on loading into R, organizing and pre-processing the spectra. Once these steps are accomplished, the spectra are ready for further exploratory analysis. In this chapter, we do not make use of ancillary laboratory soil analysis and focus only on the information that can be obtained directly from the spectra. This involves the study of the patterns and peaks present in the spectra, which can be achieved in different ways.

We first describe relatively simple methods to identify physical and chemical properties of soils directly from soil visible and near-infrared (vis-NIR) spectra. This is possible because absorption of visible and infrared energy in soils is a response of its constituent physical and chemical properties. Also exploited is the fact that certain physical and chemical properties have diagnostic responses in the vis-NIR range. With examination of a soil spectrum, one can infer the presence or absence of certain properties and to an extent their relative abundances. In this chapter, we describe specifically the detection of secondary clay minerals and iron oxides. An emphasis is given to their detection in the vis-NIR region of the spectrum, but much of the foundational work on assessing the absorbance properties of particular clays and minerals was done using mid-infrared spectra analysis (see Farmer 1974). These foundational efforts have been subject to updates and improvements, summarized in Madejova et al. (2017).

When conducting exploratory analysis of spectral data, we are immediately burdened with the issue of high dimensionality. It is such that we may be dealing with (using NIR spectra data as an example) over 2000 individual wavelengths for each spectrum. When one wants to investigate patterns in the data, spectral similarities and differences, or to detect spectral outliers, it is necessary to reduce the dimension of the spectra while keeping the important features. A natural candidate of this is principal component analysis (PCA). PCA has been used extensively in many varied fields of research. Since there are many literature contributions describing its theoretical underpinnings (e.g. Wold et al. 1987, Abdi and Williams

A. M. J.-C. Wadoux et al., *Soil Spectral Inference with R*, Progress in Soil Science,
https://doi.org/10.1007/978-3-030-64896-1_6

2010 or Varmuza and Filzmoser 2016), this chapter focuses only on applications using spectral data.

Finally, we provide examples on how to use these exploratory analyses in the two last sections of this chapter. In the first example, we define the spectral prediction domain of the existing spectral library with respect to new recorded spectra. In the second example, we derive soil colour from the spectra using specific wavelengths in the visible range of the spectrum.

The set of packages used in this chapter are installed using the lines below. The book-associated `soilspec` package is also required; see Chap. 3 for information on its installation.

```
#specify allthe packages usedin the chapterand install them if they are not already
myPackages <- c("SDMTools", "tripack", "resemble", "splancs", "sp")

# define which packages are not installed in the current computer
notInstalled <- myPackages[!(myPackages %in% installed.packages()[ , "Package"])]

# install the missing packages
if(length(notInstalled)) install.packages(notInstalled)
```

6.1 Feature Selection

6.1.1 *Identifying Secondary Clay Minerals and Iron Oxides*

Clay minerals and iron oxides absorb at specific wavelengths in the vis-NIR range of the electromagnetic spectrum (Clark et al. 1990). Previous studies have demonstrated the use of soil vis-NIR spectra in soil compositional studies with notable success (Brown et al. 2006; Viscarra-Rossel et al. 2009). The idea behind assessing the mineral composition of soils with vis-NIR spectroscopy is to compare the reflectance of the diagnostic wavelengths from a given reference spectra with the reflectance at the same wavelengths of the soil samples. The major diagnostic wavelengths for detecting some clay minerals and iron oxides are summarized in Table 6.1. A useful repository of reference material with corresponding measured spectra is the U.S. Geological Survey digital spectral library (Clark et al. 2007).

Table 6.1 Some diagnostic vis-NIR wavelength ranges of selected secondary clay minerals and iron oxides

Mineral	Diagnostic wavelength range(s)/nm
Kaolinite	2078–2267
Smectite	2118–2287
Kaolinite-smectite 50-50 mixture	2128–2258
Illite	2155–2266, 2306–2385
Goethite	457–563, 776–1266
Hematite	455–612, 765–1050

To initiate the comparison between the reference material spectrum $\mathbf{x}_{\text{ref}}$ and soil spectrum $\mathbf{x}_i$, first (for each of the reference spectrum and individual soil spectrum), the specific wavelength ranges diagnostic to each clay mineral and iron oxide specimen is isolated. Each range is then normalized using the continuum removal technique. One method of continuum removal fits a convex hull to the reflectance spectra over the diagnostic range. This convex hull provides the continuum, and the relative deviation from this continuum is used to calculate the continuum-removed spectrum (Clark and Roush 1984). With the continuum-removed spectra, the presence of each clay mineral and iron oxide in each spectrum can be estimated when it is compared to that of the continuum-removed reference spectrum. The approach is exemplified by the US Geological Survey and their Tetracorder decision-making framework for identifying minerals and inorganic compounds (Clark et al. 2003). Fundamentally, Tetracorder uses a shape-fitting algorithm, which essentially reduces down to a correlation between the reference and observed soil spectra. The correlation coefficient (r; see Sect. 9.1) is a quantitative estimate of the shape similarity between the reference and soil spectra. Other criteria used for estimating similarity to the reference spectrum are the relative band depth and relative area of the soil spectral feature (with respect to the reference). The slope of the continuum is also used, particularly for the positive identification of iron oxides.

The band depth of a spectral feature for a given clay mineral or iron oxide is defined by (Clark and Roush 1984):

$$d = 1 - \frac{s_{cr}}{s_l} \tag{6.1}$$

where d is the band depth of the spectral feature, s_{cr} is the reflectance value at the minimum of the continuum-removed spectrum of $\mathbf{x}_{\text{ref}}$ for the diagnostic wavelength range and s_l is the reflectance value of the continuum line at the same wavelength. A relative depth d' can be calculated as the ratio of the band depth of an unknown soil sample to that of a reference spectrum. The area between the continuum-removed spectrum of $\mathbf{x}_{\text{ref}}$ for the diagnostic wavelength range and continuum line at the same range is estimated by the conventional area calculation method. The relative area a' between the area of a soil spectrum and the area of a reference spectrum is calculated as the ratio between the two. Finally, the relative abundance of a clay mineral or iron oxide in a particular soil sample can be derived by:

$$\text{relative abundance} = r \times d' \times a', \tag{6.2}$$

where d' is the relative depth of the spectral feature for the diagnostic wavelength for a given reference mineral or iron oxide and a' is the relative spectral feature area. In the case where there is more than one diagnostic spectral feature such as for illite or both the iron oxide hematite and goethite, the relative abundance is derived by:

$$\text{relative abundance} = \sum_{i=1}^{nb} c_i \times r_i \times d_i' \times a_i', \tag{6.3}$$

where nb is the number of diagnostic spectral features and c_i is the proportional area of the reference spectral feature i to the total summed area of the (reference mineral) spectral features.

To put what has just been described into a more practical learning, the following exercise will demonstrate:

1. How to use one of the reference material spectra extracted from the USGS repository and compute the continuum removal of this spectrum within the specified diagnostic wavelengths.
2. How to use this continuum-removed reference material spectrum to estimate a number of parameters (shape, area, depth, slope) that quantitatively describe absorption features within the diagnostic wavelength range.
3. How to estimate the relative abundance of clay minerals and iron oxides from your own spectra.

The USGS spectral library is a database of spectra together with sample descriptions of many hundreds of samples and reference materials. We have extracted and post-processed a few of these spectra that correspond to the clay minerals and iron oxides mentioned previously. These are collectively contained in the file `mineralRef` from the book package `soilspec`.

```
# load the book package
require(soilspec)

# load the reference spectra
data("mineralRef")

# inspect the object
str(mineralRef)
```

```
## 'data.frame':    2151 obs. of  13 variables:
##  $ wavelength          : int  350 351 352 353 354 355 356 357 358 359 ...
##  $ Geothite            : num  0.0191 0.0191 0.0189 0.0187 0.0185 0.0183 0.0181 0.018
                                 0.0179 0.0178 ...
##  $ hematite            : num  0.0199 0.019 0.0182 0.0176 0.0172 0.017 0.017 0.0171
                                 0.0172 0.0174 ...
##  $ gypsum              : num  0.872 0.873 0.874 0.874 0.874 ...
##  $ calcite             : num  0.796 0.798 0.799 0.799 0.798 ...
##  $ kaolinite_114       : num  0.324 0.329 0.334 0.339 0.343 ...
##  $ kaolinite_113       : num  0.386 0.392 0.399 0.405 0.411 ...
##  $ montmorillinite_126: num  0.32 0.322 0.324 0.325 0.327 ...
##  $ monmorillinite_127 : num  0.502 0.504 0.505 0.507 0.508 ...
##  $ illite_121          : num  0.355 0.356 0.358 0.359 0.36 ...
##  $ illite_120          : num  0.154 0.154 0.153 0.153 0.153 ...
##  $ kaol_smect_124      : num  0.114 0.116 0.118 0.121 0.123 ...
##  $ kaol_smect_125      : num  0.136 0.138 0.14 0.142 0.144 ...
```

You will note that for some minerals there is more than one spectrum. In fact, as you will note from the USGS spectral library, there can be many tens of samples of the same reference material. They may differ in terms of provenance and grain size. Collectively these differences will result in changes to the absorbance features within the diagnostic wavelengths. The actual range of the diagnostic wavelengths may differ slightly from reference material to reference material too. This does not present a serious problem because when a sample is presented to a diagnostic engine such as the USGS Tetracorder (as an example), the presented sample spectrum is compared to the whole spectral library where the comparison criteria are estimated for every sample in the library. It is such that one reference material may indicate absence of a specified clay mineral, while a different sample of the same type of reference material may indicate that the clay mineral is in abundance. If there is a high likelihood that a clay mineral is in abundance from at least one of the reference materials, it would be considered as confirmatory that clay mineral is in fact present in the undiagnosed sample being evaluated.

As a first step, we plot the two spectra that correspond to the kaolinite reference material. An identifying characteristic of kaolinite is the doublet spectra feature in the 2200 nm region. A similar doublet feature is also present in the 1400 nm region. The absorption feature near 1400 nm is due to overtones of the O-H stretch vibration near 2778 nm (3600 cm^{-1}), while those near 2200 nm are due to Al-OH bend plus O-H stretch combinations (Stenberg et al. 2010). Subsequently water or soil moisture could have a stronger attenuating effect on the spectra in the 1400 nm region compared to 2200 nm region because of these different overtone properties.

```
# create a sequence of numbers to represent the wavelength
wavelength <- seq(350, 2500, by = 1)

# plot the kaolinite spectrum number 114
plot(wavelength, (t(mineralRef$kaolinite_114)),
     type = "l",
     col = "blue",
     ylim = c(0.2, 1.2),
     xlab = "Wavelength /nm",
     ylab = "Reflectance")

# add to the plot the kaolinite spectrum number 113
lines(wavelength, (t(mineralRef$kaolinite_113)),
      type = "l",
      col = "red")

# add a legend
legend("topright",
       legend = c("kaolinite_114", "kaolinite_113"),
       lty = c(1, 1),
       col = c("blue", "red"))
```

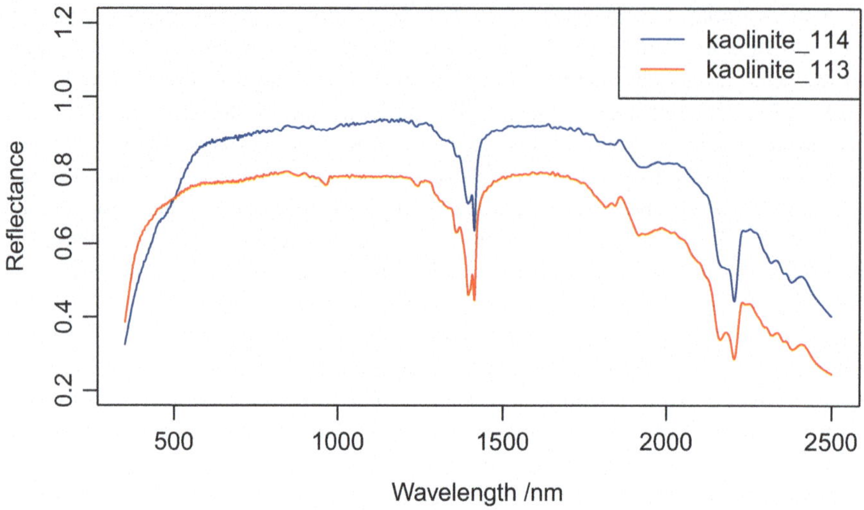

Fig. 6.1 Example of two reflectance reference spectra for kaolinite, called `kaolinite_114` and `kaolinite_113` in the USGS spectral library

Using the diagnostic wavelengths indicated for kaolinite in Fig. 6.1, we can trim the spectrum accordingly using the `trimSpec` function defined in Sect. 5.1 (Fig. 6.2).

```
# select reference mineral
kaolRef <- as.data.frame(t(mineralRef$kaolinite_114))
colnames(kaolRef) <- wavelength

# diagnostic wavelength range
lower <- 2078
upper <- 2267
```

```
# spectrum trimming
kaolDiog1 <- trimSpec(kaolRef,
                      wavlimits = c(lower,upper))

# plot the trimmed reflectance spectrum between the wavelengths 2078 and 2267nm
plot(colnames(kaolDiog1[1,]), kaolDiog1[1,],
     type = "l",
     xlab = "Wavelength /nm",
     ylab = "Reflectance")
```

Next we use a continuum removal procedure to normalize the spectra. The function to be used here is called `chBLCext` from the book package `soilspec`. In addition to returning the continuum-removed spectra, the `chBLCext` function also returns the fitted continuum, plus the vertices needed for calculating the area of the

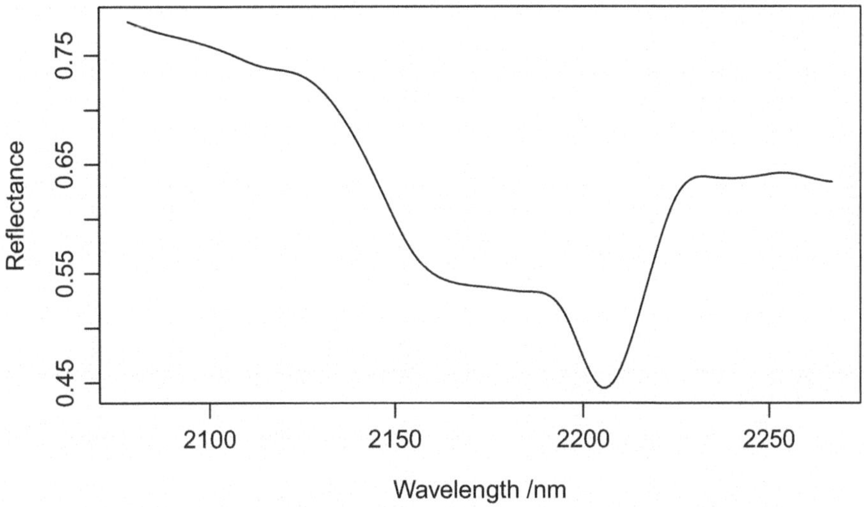

Fig. 6.2 Trimmed spectra for the region of interest of reflectance wavelengths for kaolinite

diagnostic spectral feature. Essentially, we need to input the spectra, together with the `upper` and `lower` bounds of the diagnostic range. The `type` argument needs also to be specified, to tell the function if we are providing either an absorbance or reflectance spectrum.

We can apply this function to the reference spectra (Fig. 6.3).

```
# continuum removal function
kaolDiog1CR <- chBLCext(spectra = kaolDiog1,
                        wav = as.numeric(colnames(kaolDiog1)),
                        type = "R")

# plot the trimmed reflectance spectrum between the wavelengths 2078 and 2267nm
plot(colnames(kaolDiog1[1,]), kaolDiog1[1,],
     type = "l",
     xlab = "Wavelength /nm",
     ylab = "Reflectance")

# add the convex hull derived continuum
lines(colnames(kaolDiog1[1,]), kaolDiog1CR$continuum,
      col = "red")
```

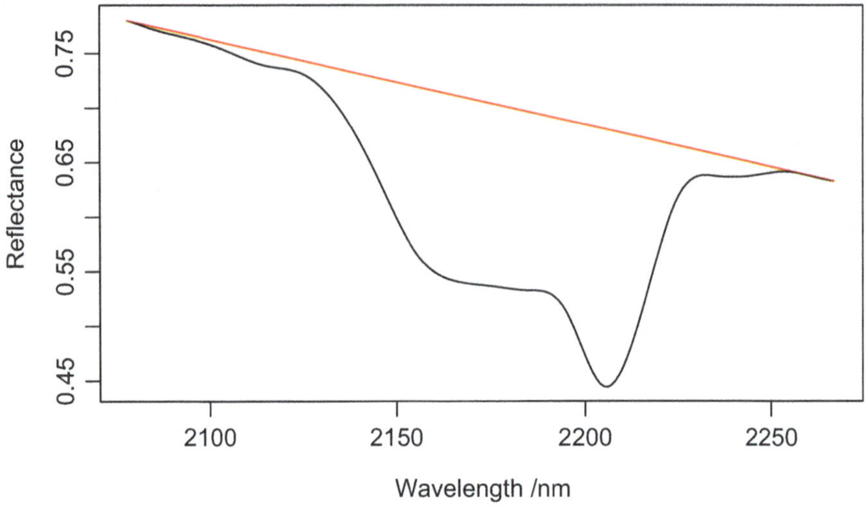

Fig. 6.3 Continuum fitted to reference material spectrum for the diagnostic wavelength range of `kaolinite_113`

We can then plot the continuum-removed spectrum which is calculated as $\frac{reflectance}{continuum}$ across the diagnostic wavelength range.

Continuum fitted to reference material spectrum for the diagnostic wavelength range of `kaolinite_113`. Continuum-removed spectrum for the diagnostic wavelength range of `kaolinite_113` (Fig. 6.4).

```
# plot continuum-removed spectra
plot(colnames(kaolDiog1[1,]), kaolDiog1CR$cHull,
     type = "l",
     xlab = "Wavelength /nm",
     ylab = "Continuum-removed reflectance")
```

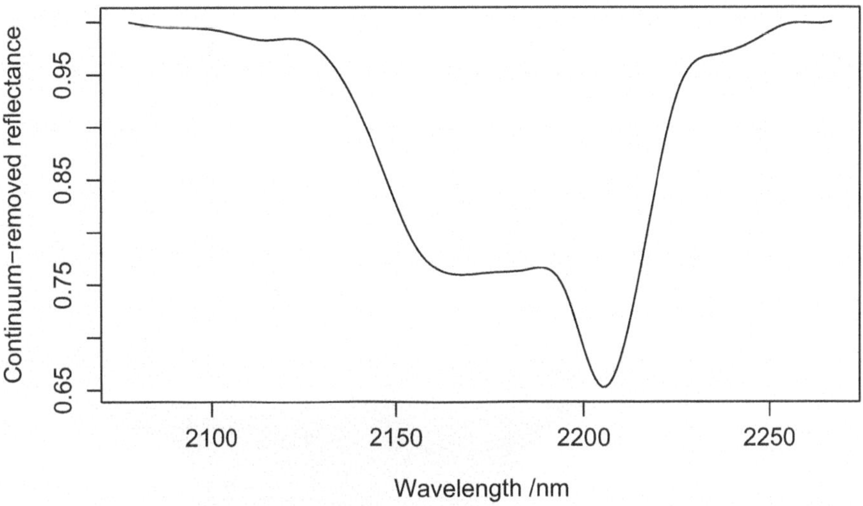

Fig. 6.4 Continuum-removed spectrum for the diagnostic wavelength range of `kaolinite_113`

Quantitative parameters of absorbance features

Now we wish to estimate specifically the absorbance feature's depth, area and slope. Following from Equation (6.1), the estimation of band depth can be scripted as below. All we are doing here is searching for the minimum reflectance value of the continuum-removed spectrum and then searching for the corresponding value (s_{cr}) of the raw reflectance and the convex hull continuum (s_l) at the same wavelength.

```
# find wavelength corresponding to minimum continuum-removed reflectance
waveId <- which(kaolDiog1CR$cHull==min(kaolDiog1CR$cHul))[1]

# identify raw reflectance reading
Scr <- kaolDiog1CR$rawSpec[waveId]

# identify continuum reading
Sl <- kaolDiog1CR$continuum[waveId]

# calculate band depth, Eq. 7.1
featureDepth <- 1 - (Scr/Sl)

# display band depth
featureDepth
```

```
##          2205
## 1 0.3474705
```

Estimating the area of the polygon is relatively straightforward. The vertices of the spectral feature have been stored as an element within the `kaolDiog1CR` object. We can configure these vertices to make a polygon after which we calculate the area. For the area calculation, we use the `areapl` function from the `splancs` package (Fig. 6.5).

```
# plot continuum-removed spectra
plot(colnames(kaolDiog1[1,]), kaolDiog1CR$cHull,
     type = "l",
     xlab = "Wavelength /nm",
     ylab = "Continuum-removed reflectance")

# polygon covering the absorbance feature
polygon(kaolDiog1CR$polygon,
        col = "red",
        border = NA)
```

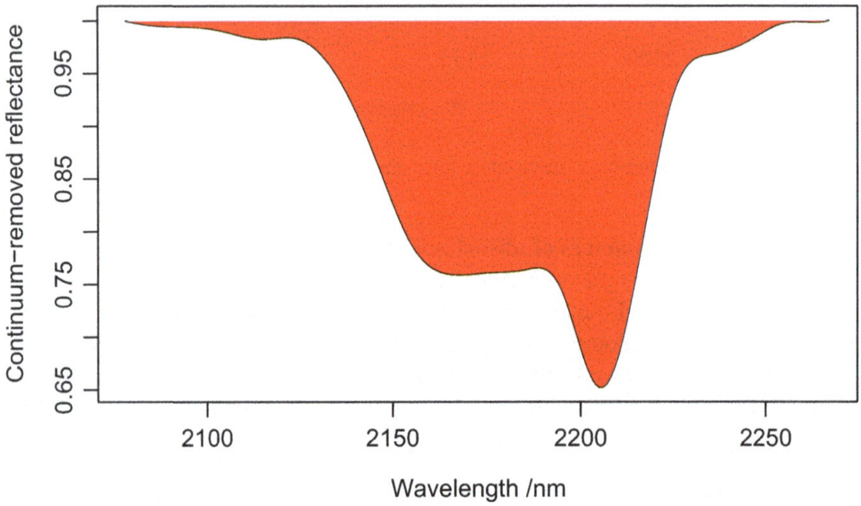

Fig. 6.5 Area (in red) of the absorbance feature of the kaolinite for the region of interest

```
# load the required package
require(splancs)

# calculate area of the polygon
featureArea <- areapl(kaolDiog1CR$polygon)

# display area of the polygon
featureArea
```

```
## [1] 21.23712
```

Finally calculating the slope of the continuum is just the value of the continuum at the smallest wavelength subtracted from the value of the continuum at the largest wavelength.

```
# calculate slope of the continuum
featureSlope <- tail(kaolDiog1CR$continuum, 1) - kaolDiog1CR$continuum[1]

# display value of the slope
featureSlope
```

```
## [1] -0.1471
```

For future reference and for the later comparison, it is handy to store all the pertinent information that has been gathered regarding each reference material as one object. Below all the information we have gathered about the kaolinite reference material are stored in a `list` object called `kaolin1141summary`. This is effectively the procedure we have followed for all the other reference materials to build a library of summaries that we use in the next section to analyse some soil spectra.

```
# save a summary of all the information gathered about the reference material
kaolin1141summary <- list(name = "kaolinite114material",
                          wave = as.numeric(colnames(kaolDiog1[1,])),
                          CRspectra = kaolDiog1CR$cHull,
                          bandDepth = featureDepth,
                          featureArea = featureArea,
                          featureSlope = featureSlope,
                          continuum = kaolDiog1CR$continuum,
                          contiuumPolygon = kaolDiog1CR$polygon,
                          rawSpectrum = kaolDiog1CR$wave)
```

6.1.2 Comparing Soil Spectra with Spectra of Reference Materials

This section is dedicated to implementing the task of comparing diagnostic spectral features of the reference material to the corresponding features of collected soil spectra. The previously saved `kaolin1141summary` contains the objects or reference material summaries of the kaolinite soil clay mineral described earlier. The process of creating these summaries is described previously. For the comparative work, we load in an undiagnosed set of soil spectra. The context of these spectra collection is described in Chap. 3.

```
# load the unknown soil spectra
data("rutherglenNIR")
```

Then we do some curation to prepare these spectra for analysis. It is at this stage where we decide whether we implement spectral smoothing of the raw spectra or not. In the following example, this option is not taken.

```
# curation
nc <- ncol(rutherglenNIR)

# remove first column (labels)
rutherglenNIR <- rutherglenNIR[, 2:nc]

# wavelength sequence
wavelength <- seq(350, 2500, by = 1)

# append column names
colnames(rutherglenNIR) <- wavelength
```

It is possible to automate the procedure of checking each sample for the presence or relative abundance of each reference material that we have summary information for. In fact, this would be a good advanced exercise to try and implement in R.

The following is an example of checking, one spectrum at a time, the relative abundance compared with a single selected reference material. In this case, the selected material is one of the kaolinite samples. The example script below however is flexible. We can change the reference material and soil spectrum to which we make the comparison.

```
# select the reference material you want to compare soil spectra too
minSelect <- kaolin1141summary

# region of Interest
lower <- min(minSelect$wave)
upper <- max(minSelect$wave)

# select the spectrum
speczz <- 1
```

Spectral trimming is next. The next script using the `specTrim` function given in Chap. 5 will actually perform the trimming upon the whole spectral dataset in one go (Fig. 6.6).

```
# spectral trimming
specTrim <- trimSpec(spectra = rutherglenNIR,
                     wavlimits = range(lower:upper))

# plot single spectrum
plot(colnames(specTrim), specTrim[speczz,],
     type = "l",
     xlab = "Wavelength /nm",
     ylab = "Reflectance")
```

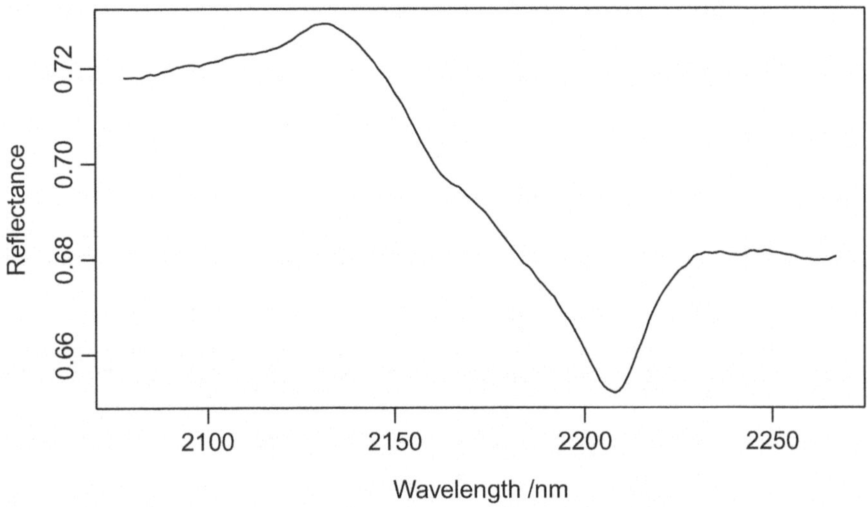

Fig. 6.6 First reflectance spectrum of the Rutherglen subset of data from the `soilspec` book package for the region of interest between 2079 and 2267 nm

Then we can fit the continuum to this spectrum using the `chBLCext` function (Fig. 6.7). This function takes as input a matrix containing the reflectance spectra and the lower and upper limits of the diagnostic ranges. In the example below, this range corresponds to kaolinite.

```
# convex Hull and continuum
specCR <- chBLCext(specTrim[speczz,],
                   wav = as.numeric(colnames(specTrim)),
                   type = "R")

# plot spectrum for the region of interest
plot(colnames(specTrim[speczz,]), specTrim[speczz,],
     type = "l",
     xlab = "Wavelength /nm",
     ylab = "Reflectance")

# add line of the continuum fitted
lines(colnames(specTrim[speczz,]), specCR$continuum,
      col="red")
```

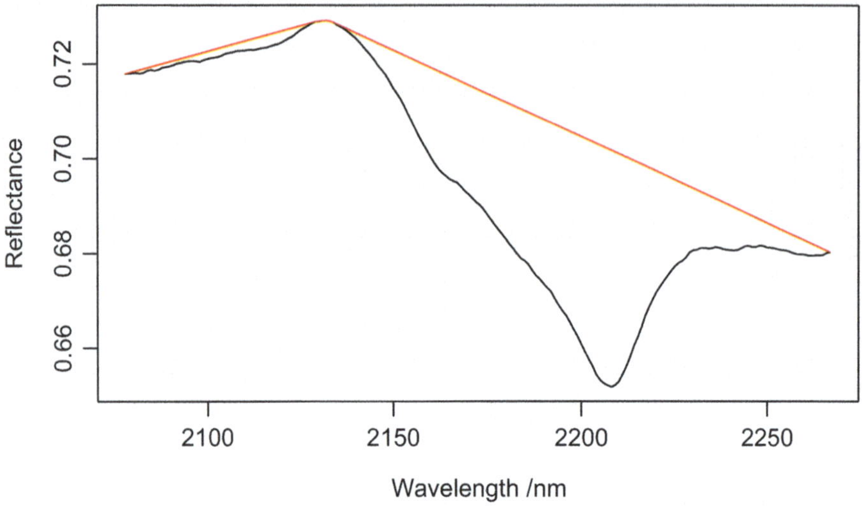

Fig. 6.7 Continuum fitted to reference material spectrum for the region of interest of reflectance bands for kaolinite

We can then calculate the various quantitative parameters that describe the absorbance features within this spectral range.

```
# band depth
sCr <- specCR$rawSpec[which(specCR$cHull==min(specCR$cHull))]
sL <- specCR$continuum[which(specCR$cHull==min(specCR$cHull))]
specd <- 1 - (sCr/sL)
specd
```

```
##          2207
## 1 0.07127731
```

```
# area of spectral feature
specArea <- areapl(specCR$polygon)
specArea
```

```
## [1] 3.843102
```

```
# slope of continuum
specSlope <- specCR$continuum[length(specCR$continuum)] - specCR$continuum[1]
specSlope
```

```
## [1] -0.03755679
```

Now we can start comparing the spectral feature of this spectrum to that of the reference material. First, we estimate the correlation between the continuum-removed spectra of the reference material to the soil spectrum. This measure provides an indicator of the similarity in shape between both features.

```
# correlation
specFit <- as.numeric(cor(t(specCR$cHull), t(minSelect$CRspectra)))
specFit
```

```
## [1] 0.9048102
```

Then we calculate the relativity parameters to spectral depth and area of the reference material. We can then also estimate relative abundance of the target mineral by applying Equation (6.3).

```
# relative depth
relativeDepth <- as.numeric(specd/minSelect$bandDepth)
relativeDepth
```

```
## [1] 0.205132
```

```
# relative area
relativeArea <- as.numeric(specArea/minSelect$featureArea)
relativeArea
```

```
## [1] 0.1809616
```

```
# fit x relative depth
specFit*relativeDepth
```

```
## [1] 0.1856055
```

```
# fit x depth x area
specFit*relativeDepth*relativeArea
```

```
## [1] 0.03358746
```

The relative abundance of kaolinite in this sample seems small, although there is quite a strong correlation between the spectral absorbance shapes in this region. A helpful plot is to overlay the continuum-removed plot of the soil spectrum upon the reference material. This plot confirms the low abundance. However, it is important to reiterate that the abundance values are not useful on their own. Abundance is not a quantitative estimate of concentration per se. However, abundance values are useful when assessing relative abundances among a group of minerals and when comparing values (of a given metric and combination thereof) between different samples (Fig. 6.8).

```
# plot reference material for the specific wavelengths
plot(minSelect$wave, minSelect$CRspectra,
     type = "l",
     col = "blue",
     ylab = "Continuum-removed reflectance",
     xlab = "Wavelength /nm")

# add to the plot the spectrum of the sample with low kaolinite content
lines(specCR$wave, specCR$cHull,
      col = "red")
```

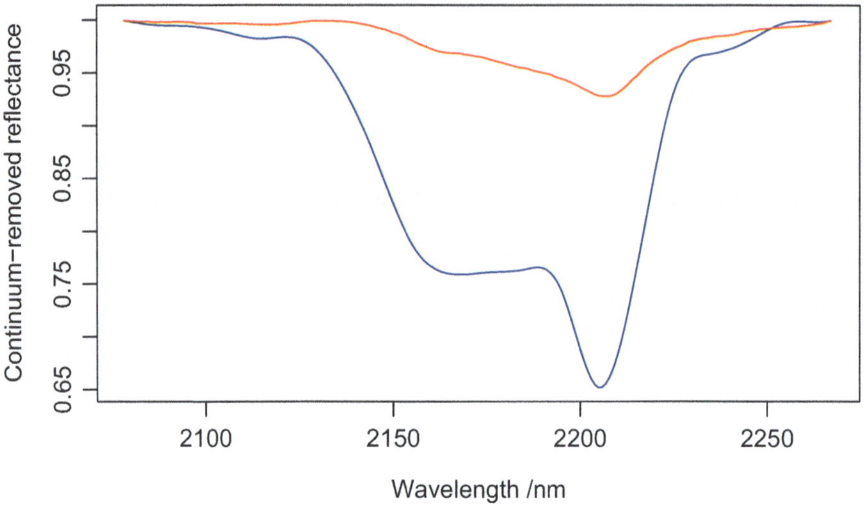

Fig. 6.8 Reference (blue) spectrum for kaolinite and example (red) spectrum in the region of kaolinite absorbance. The example spectrum shows relatively low content of kaolinite compared to the reference spectrum

In a spatial context, the estimation of relative abundances provides the opportunity to assess their variability across landscapes. Understanding the spatial variability of secondary clay minerals via the tools of digital soil mapping (McBratney et al. 2003) provides an understanding of the soil behaviour and function in a given environment. Some prior examples of such work includes Viscarra-Rossel (2011) who digitally mapped the relative abundances and distributions of kaolinite, illite and smectite in Australian soil. This work continues to inform and improve ongoing efforts to develop a consistent digital soil infrastructure for Australia (Grundy et al. 2015). At a much smaller spatial extent, Malone et al. (2014) digitally mapped secondary clay minerals and iron oxides across the Lower Hunter Valley, Australia, to aid in the much wider analysis of distinguishing viticulture terroirs in that region.

The feature identification procedures described previously are very much generalizable and potentially applied to other spectral data applications and contexts. In an interesting study, Ng et al. (2017) used the feature selection approach to select regions in MIR and NIR spectra that help quantify soil contaminants, specifically total recoverable hydrocarbons (TRHs). The authors found that predictive models of TRH concentration could be skilfully predicted when models were trained with selected spectral regions where there is sensitivity to TRHs. Such work could potentially inform the development of custom sensors and improve the efficiency of modelling to target on specific spectral regions rather than using the full wavenumber spectral range for model calibrations.

6.2 Principal Component Analysis

Principal component analysis (PCA) is a statistical technique used to examine the interrelations among a set of variables and to identify their underlying structure. In a PCA, the original variables are transformed into a (smaller) number of uncorrelated variables called principal components. The first principal component accounts for most of the variability in the data, and each succeeding component accounts for a sequentially decreasing amount of the remaining variability in the data. Thus with a few components, we may be able to explain most of the variation in the spectra. With a lower dimension dataset, we can perform outlier detection or select the spectra in which to perform laboratory-based chemical analysis so as to calibrate soil attribute predictive functions. In the following, we will examine how to use PCA for spectral data and examine or interpret some of the results.

In this section, we use the raw spectra `datsoilspc` provided by our book `soilspec` package (Fig. 6.9). The steps for pre-processing are explained in Chap. 5.

```
# load the required package
require(soilspec)

# load the data
data("datsoilspc")

# convert reflectance to absorbance
spectraA <- log(1/datsoilspc$spc)

# plot first spectrum
matplot(x = colnames(spectraA), y = t(spectraA),
        xlab = "Wavelength /nm",
        ylab = "Absorbance",
        ylim = c(0, 4),
        type = "l",
        lty = 1,
        col = rgb(red = 0.5, green = 0.5, blue = 0.5, alpha = 0.3))
```

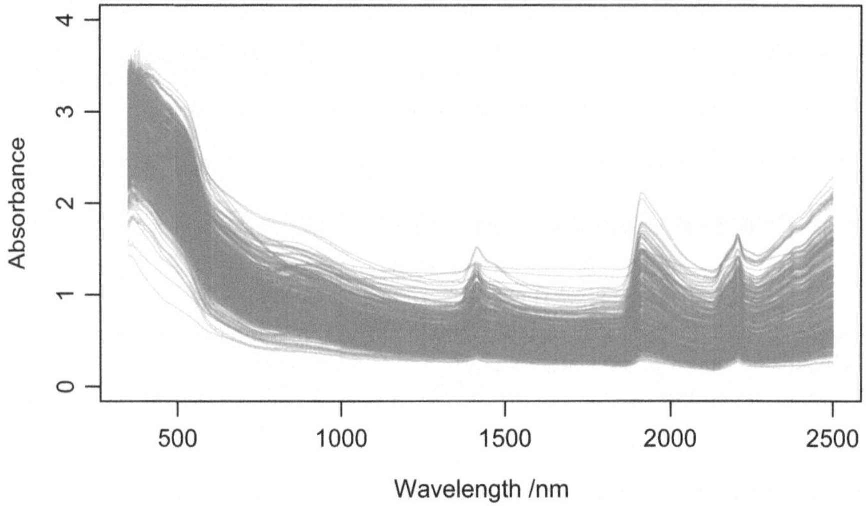

Fig. 6.9 Absorbance spectra from the book package `soilspec` dataset called `datsoilspc`

Principal component analysis (PCA) aims at approximating the matrix containing the spectra **X** by a product of two matrices, called loading and score matrices, as follows:

$$\mathbf{X} = \mathbf{T}\mathbf{P}^T, \tag{6.4}$$

where **X** is the matrix of size $n \times b$ containing the spectra, where n is the number of spectra and b is the number of spectral bands (wavelengths); $\mathbf{T} = \mathbf{XP}$ is the matrix of scores (projection of **X**) of size $n \times d$, where d is the number of principal components; and **P** is the loading matrix of size $d \times b$. The superscript T means the transpose of the matrix. PCA aims to solve for matrix **P** so that the scores $\mathbf{t}_1, \ldots, \mathbf{t}_d$ are arranged in a decreasing variance order and uncorrelated. The use of the loadings and scores is explained later in this section.

PCA in R can be implemented through the `prcomp` function which is part of the base `stats` package. This function essentially requires the spectra table, which in our case is the absorbance spectra. Several parameters may be defined, for which the most important are whether you want to centre and/or standardize the spectra. Centring means adjusting each of the spectra so that they are zero centred. Standardizing means adjusting the spectra to ensure they have unit variance. In this example, we do PCA in default mode which does not perform any prior centring or scaling to the spectra. All steps for spectral pre-processing are explained in more details in Chap. 5.

```
#principal component of the processed spectra
pcspectra <- prcomp(spectraA)

#amount of variance explained by each component 1to10
summary(pcspectra)[[6]][,1:10]
```

```
##                             PC1      PC2      PC3      PC4       PC5       PC6
## Standard deviation     11.22897 4.205422 2.407908 1.249613 0.6380026 0.5032785
## Proportion of Variance  0.82859 0.116220 0.038100 0.010260 0.0026700 0.0016600
## Cumulative Proportion   0.82859 0.944810 0.982910 0.993170 0.9958500 0.9975100
##                              PC7       PC8       PC9      PC10
## Standard deviation     0.3910278 0.2012169 0.1678252 0.1480202
## Proportion of Variance 0.0010000 0.0002700 0.0001900 0.0001400
## Cumulative Proportion  0.9985200 0.9987800 0.9989700 0.9991100
```

Alternatively, the PCA can be performed with the `pc_projection` function from the `resemble` package. In this case, the user must first decide what is the maximum amount of cumulative variance that needs to be retained by the PCs. In the next chapters, we will use both the `base` function `prcomp` and the `resemble` function `pc_projection` interchangeably.

```
# load the required package
require(resemble)

# specify amount of cumulative variance that needs to be retained by the PCs.
# note that in most cases 0.99 is sufficient
maxexplvar <- 0.99

# perform the PCA
pcspectraRs <- pc_projection(Xr = spectraA,
                             pc_selection = list("cumvar", maxexplvar),
                             method = "pca",
                             center = TRUE, scale = FALSE)

# display the information contained in the pcspectraRs object
print(pcspectraRs)
```

```
##
##  Method:  pca (svd)
##  Number of components retained:  3
##  Number of observations and number of original variables:  391 2151
##
##  Standard deviations, cumulative variance explained, individual variance explained:
##
##  Explained variance in X {Xr; Xu}:
##                            pc_1  pc_2   pc_3
## sd                       11.229 4.205 2.4079
## cumulative_explained_var  0.829 0.945 0.9829
## explained_var             0.829 0.116 0.0381
```

It is possible to plot the amount of variance explained for each of the principal component using the `screeplot` function (Fig. 6.10).

```
# plot the variances explained by each component
screeplot(pcspectra,
          npcs = 10,
          main = " ",
          type = "lines")
```

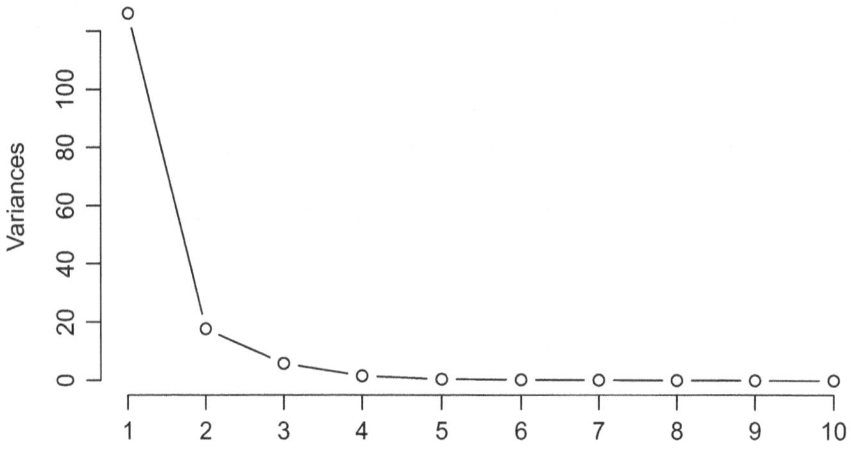

Fig. 6.10 Amount of variance retained for each of the principal components. Only the ten first components are displayed

The scree plot illustrates the amount of variance in the spectra that is described by each component. It is always the case that the first component explains most of the variation. Each successive component decreasingly explains a little bit less of the variation. Here the first five components describe over 99% of the spectral variation. Alone the first component describes 83%. We have reduced what was a very high dimensional spectral dataset down to just a few components.

Biplots are a useful output of PCA, where they visually show both individual points and the variables. PCA scores are the original spectra in a rotated coordinate system. Relation between scores and the original data are determined via the loadings. The loadings can be understood as the weights for each original variable (here wavelength) when calculating the principal component. By multiplying the transpose of the variable loadings by the PCA scores, the original spectra table will be returned. By plotting the PCA scores, we can evaluate visually any relationships among the different spectra and wavelengths, for example, whether there is clustering or grouping among the spectra. The loadings are used to interpret relationships between variables, which is aided by the use of arrows on the plot. The length of the lines approximates the variances of our variables (the wavelengths). The longer the line, the higher the variance. The angle between the lines (actually cosine of the angle) approximates the correlations between the variables. The closer the angle is to 90 or 270 degrees, the smaller the correlation. An angle of 0 or 180 degrees reflects a correlation of 1 or −1, respectively. By plotting the biplot of our first two principal components, we see (Fig. 6.11):

```
# make a biplot of the first two principal components scores and loadings
biplot(pcspectra,
       col = c(rgb(red = 0.5, green = 0.5, blue = 0.5, alpha = 1),
               rgb(red = 1, green = 0, blue = 0, alpha = 0.5)),
       xlab = "PC 1",
       ylab = "PC 2")
```

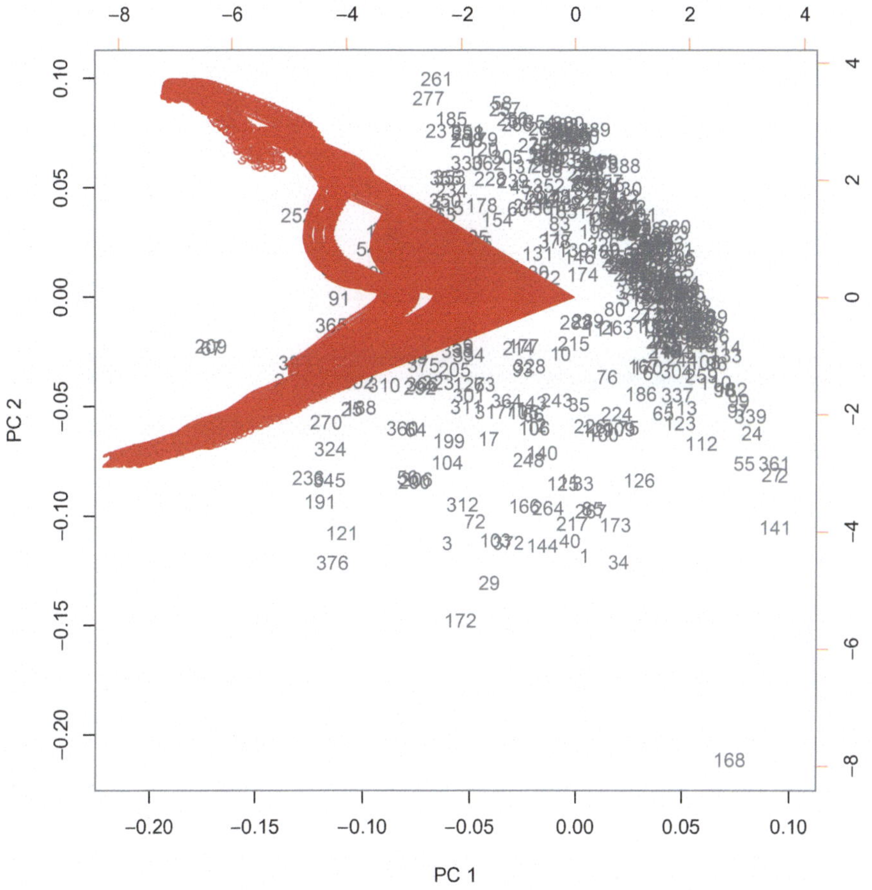

Fig. 6.11 Biplot of the first two principal components of the `datsoilspc` data. The grey numbers indicate the location of the score values in the principal component space, and the red numbers indicate the loadings

This plot is a little difficult to interpret because there are so many variables (wavelengths) under consideration, but the numbers (in grey) are the individual spectra (projected onto the first two components), while the arrows are the loadings. Many of the wavelengths are correlated with each other, and some are more variable than others. Similarly, there do appear to be some groupings within our spectra collection, and there also appear to be some spectra that could be outliers too (we will investigate this Chap. 7).

Individually we may just first want to look at the loadings plot of the first two principal components which as we know collectively explains just over 94% of the spectral variance:

```
# plot the first loading of the PC of the absorbance spectra
plot(colnames(spectraA), pcspectra$rotation[,1],
     type ="l",
     ylab ="Loading",
     xlab ="Wavelength /nm",
     col = rgb(red = 1, green = 0, blue = 0, alpha = 1),
     ylim = c(-0.06, 0.06))

# add to the plot the second loading of the PC of the absorbance spectra
lines(colnames(spectraA), pcspectra$rotation[,2],
      type ="l",
      ylab ="Loading",
      xlab ="Wavelength /nm",
      col = rgb(red = 0, green = 0, blue = 1, alpha = 1))

# add a legend
legend("topright",
       legend = c("First loading", "Second loading"),
       lty = c(1, 1),
       col = c("red", "blue"))
```

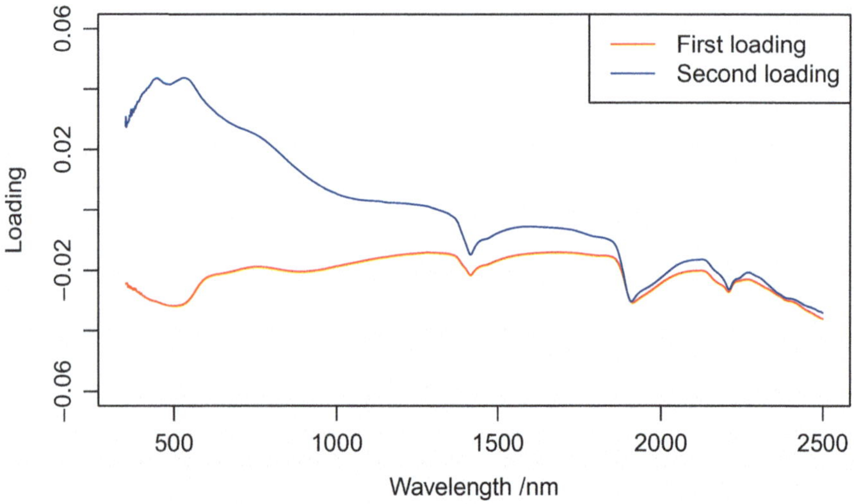

Fig. 6.12 First two loadings of the principal components of the absorbance spectra

Figure 6.12 shows the value of the coefficients for each wavelength and each of the principal component. Red and blue lines indicate the first and second principal component loadings, respectively. These coefficients, when multiplied by the PCA scores, recreate the original spectra. Each of the loadings describes a different part of the spectral structure. We can compare these loadings with that of the last or

389th principal component. With this principal component, essentially there is little to interpret, and there are no discernible patterns or features in the loading spectrum. The signal is just describing noise (Fig. 6.13).

```
# plot last principal component loading
plot(colnames(spectraA), pcspectra$rotation[,ncol(pcspectra$rotation)],
     type = "l",
     ylab = "Loading",
     xlab = "Wavelength /nm",
     col = rgb(red = 0.5, green = 0.5, blue = 0.5, alpha = 0.5),
     ylim = c(-0.1, 0.1))
```

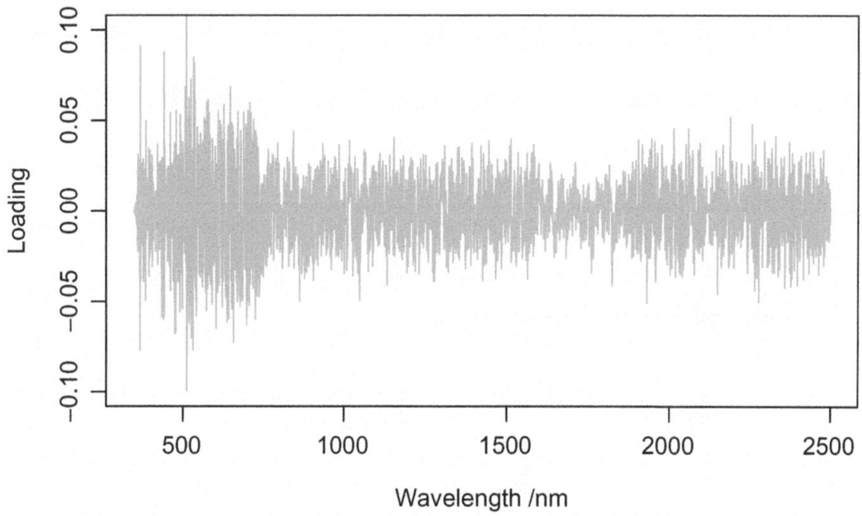

Fig. 6.13 Last (391st) loading of the principal components of the absorbance spectra

Similarly, we can plot the scores of the first two principal components, which display what was shown on Fig. 6.11 but without the loadings (Fig. 6.14).

```
# principal component scores
pcsscores <- pcspectra$x
plot(x = pcsscores[,1],
     y = pcsscores[,2],
     xlab ="PCA 1",
     ylab ="PCA 2",
     pch = 16,
     col = rgb(red = 0.5, green = 0.5, blue = 0.5, alpha = 0.5))
```

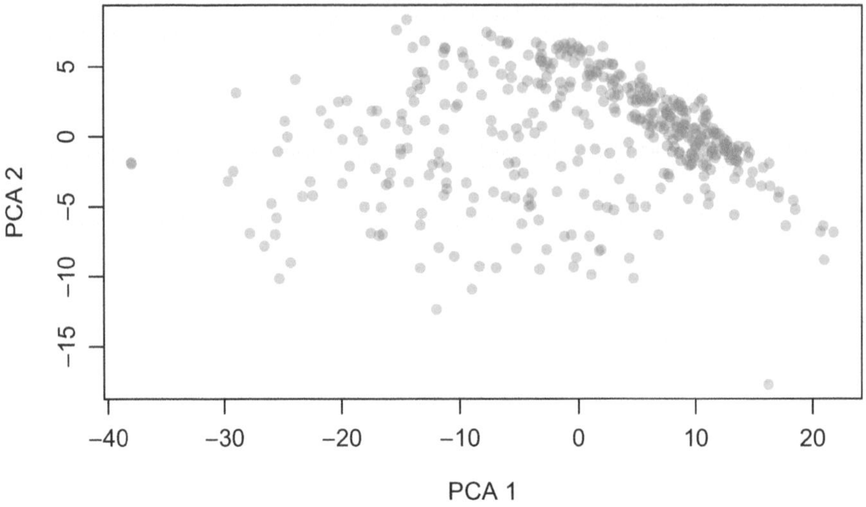

Fig. 6.14 First two principal component scores of the absorbance spectra

6.3 Spectral Prediction Domain

In our example, we know that the first two principal components describe much of the variability (94.5%) of the entire spectra. If we define a convex hull around these points, we are effectively defining a domain of prediction. It is possible to evaluate whether a prediction model based on our spectral library can realistically infer soil attribute values on new, exogenous and undiagnosed spectra. The new spectra can be projected onto the first two principal components of the original spectral library. If the points lie inside the convex hull of our spectral library, we can be confident that the spectral library and the new spectra have similar characteristics. On the other hand, if the projected spectra are outside this convex hull, it indicates that the new spectra are unlike the others in the library and as such, making conclusions about the new spectra based on the original spectral library is to be made with caution.

Using the `tri.mesh` and `convex.hull` functions from the `tripack` package, we can fit the convex hull around the PCA scores of the first two principal components. We can then visualize this as below (Fig. 6.15):

```
# load the required package
require(tripack)

# perform a triangulation of the PCA scores
randTr <- tri.mesh(pcsscores[,1], pcsscores[,2])

# from triangulation returns the nodes that are on boundary i.e. convex hull
randCh <- convex.hull(randTr, plot.it = F)

# save the convex hull vertices
prPoly = cbind(x = c(randCh$x), y = c(randCh$y))
```

```
# plot PCA scores
plot(pcsscores[,1], pcsscores[,2],
     xlab ="PCA 1",
     ylab ="PCA 2",
     xlim = c(min(pcsscores[,1:2]), max(pcsscores[,1:2])),
     ylim = c(min(pcsscores[,1:2]), max(pcsscores[,1:2])),
     pch = 16,
     col = rgb(red = 0.5, green = 0.5, blue = 0.5, alpha = 0.5))

# draw convex hull
lines(c(randCh$x, randCh$x[1]), c(randCh$y,randCh$y[1]),
      col="red",
      lwd=1)
```

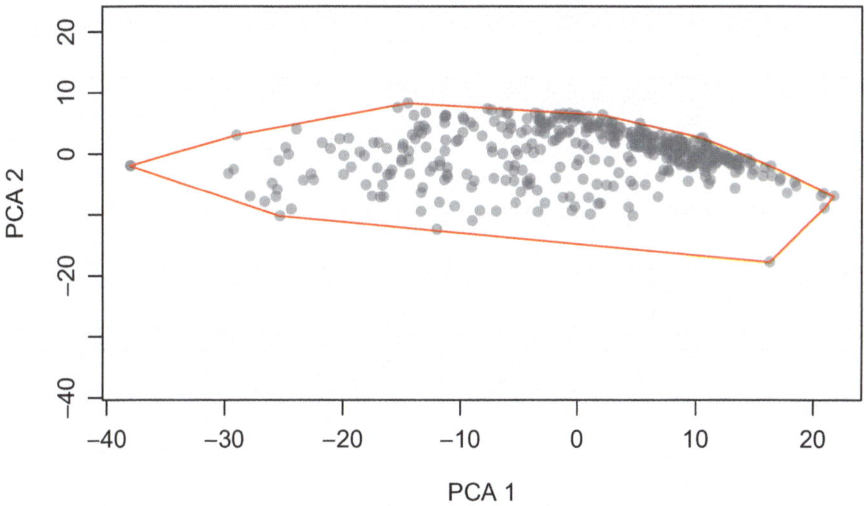

Fig. 6.15 First two principal component scores of the absorbance spectra (grey dots) and fitted convex hull (red line)

We have now defined our domain of prediction, and it is clear from the plot that there is a grouping of spectra around the high score values for PCA 1 and relatively high score values for PCA 2. For now, we will not take this further. Later on in this book, we will use this domain of prediction to determine whether a particular spectral library is appropriate for prediction of soil properties given some exogenous or introduced spectra.

One can import some new spectra and determine whether they fit inside the domain of prediction of our spectral library. The exogenous spectra are raw NIR spectra collected using the same type of instrument which the spectra making up our spectral library were collected with. After importing into R, we then need to carry out the same pretreatments as we did for our spectral library and then project them onto the principal components. We can then implement a simple procedure whereby we determine whether each exogenous spectrum fits inside the convex hull or prediction domain of our spectral library.

For the example, we modify the current spectra to simulate exogenous spectra. In a real-world case study, the exogenous spectra often come from a different case study (Fig. 6.16).

```
# set the seed to make a reproducible example
set.seed(19101991)

# create a function to apply a systematic shift to each spectrum
FUN <- function(x){
  # add a random systematic shift to each spectra
  x <- x + rnorm(1, 0, 0.1)
  return(x)
}

# apply the function to each spectra
spectraAMod <- apply(spectraA, 1, FUN)

# plot original spectrum
matplot(x = colnames(spectraA), y = t(spectraA)[,1:10],
        xlab = "Wavelength /nm",
        ylab = "Absorbance",
        ylim = c(-0.1, 4),
        type = "l",
        lty = 1,
        col = rgb(red = 0.5, green = 0.5, blue = 0.5, alpha = 0.3))

# plot modified 'exogenous' spectra
matlines(x = colnames(spectraA), y = spectraAMod[,1:10],
         xlab = "Wavelength /nm",
         ylab = "Absorbance",
         type = "l",
         lty = 1,
         col = rgb(red = 1, green = 0, blue = 0, alpha = 0.3))
```

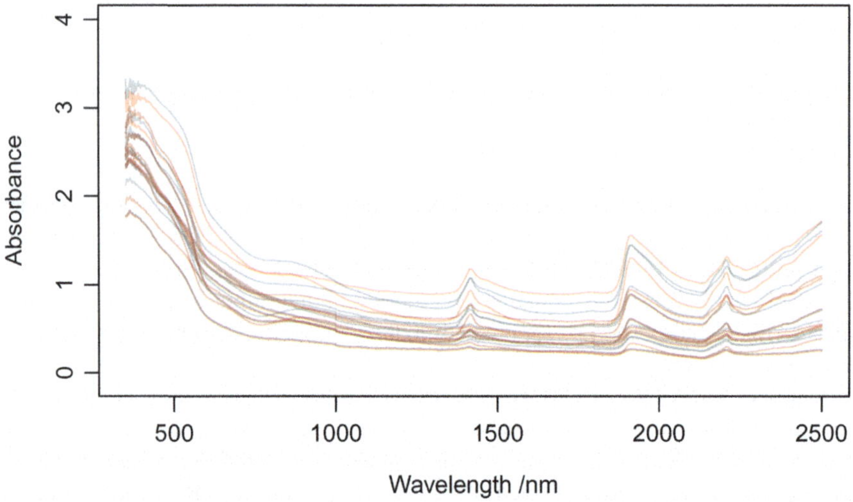

Fig. 6.16 Original (grey) and modified (exogenous, in red) first ten absorbance spectra of the `datsoilspc` data

Then we project the pretreated exogenous spectra onto the principal components that we derived for our original spectral library. This is done using the `predict` function.

```
# project the exogenous spectra onto the PC space of the original spectra
PCAProjection <- predict(pcspectra, t(spectraAMod))
```

Then we plot the PCA scores of the first two principal components for our exogenous spectra and see whether they fit within the convex hull that was created previously. Key to this procedure is a function from `SDMTools` called `pnt.in.poly` which as suggested works out if points lie within the boundaries of a defined polygon (Fig. 6.17).

```
# load required package
require(SDMTools)

# check whether the spectra fit in the polygon PCA scores of exogenous  spectra
newScores = cbind(x = PCAProjection[,1], y = PCAProjection[,2])

# plot the polygon and all points to be checked
plot(newScores,
     xlab ="PCA 1",
     ylab ="PCA 2",
     xlim = c(min(pcsscores[,1:2]),max(pcsscores[,1:2])),
     ylim = c(min(pcsscores[,1:2]),max(pcsscores[,1:2])),
     col= rgb(red = 0.5, green = 0.5, blue = 0.5, alpha = 0.5),
     pch = 16)
polygon(prPoly,
        col = rgb(red = 0.3, green = 0.3, blue = 0.3, alpha = 0.3))

# check which points fall within the polygon
specMatch = pnt.in.poly(newScores, prPoly)
points(specMatch[which(specMatch$pip==0),1:2],
       pch = "X",
       col = rgb(red = 1, green = 0, blue = 0, alpha = 1))
```

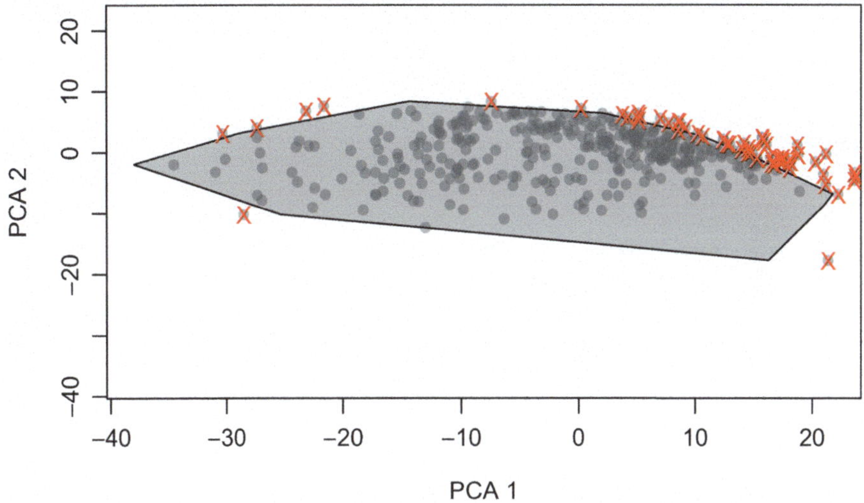

Fig. 6.17 Exogenous spectra projected onto the principal component space of the original spectra. The points lying outside the convex hull of the original spectra (grey area) are in red

Check the percentage of points falling inside the polygon.

```
# percent of points falling outside the polygon
sum(specMatch$pip)/nrow(specMatch)*100
```

```
## [1] 84.65473
```

From this plot, we can see that about 85% of our spectra fit inside the convex hull. For the remaining 15%, we should take any prediction of soil properties based on these spectra with care, as the spectral library and or its spectral prediction domain does not extend to these observations.

6.4 Soil Colour

Some soil properties can be estimated directly from the spectra without the need for further pretreatments or statistics. For example, as can be expected, the visible part of the vis-NIR spectrum contains information about colour. This information can be used to predict soil colour directly from the spectra (Mouazen et al. 2007; Summers et al. 2011).

To achieve this, we identify the average reflectance in the red (600–690 nm), green (520–600 nm) and blue (420–520 nm) colour bands. This data is used to predict colour in RGB colour space which can then be converted to Munsell notation and other colour spaces.

Let us load some data as example. They are in the book package `soilspec` and are called `specSoilCol`.

```
# load required package
require(soilspec)

# load the spectra for this example
data("specSoilCol")

# plot each spectrum separately
par(mfcol = c(3, 4))
for(i in 1:nrow(specSoilCol)){
  plot(x = colnames(specSoilCol[-c(1:2)]), y = specSoilCol[i,-(1:2)],
       type = "l",
       ylim = c(0, 1),
       main = paste0("Soil - ", specSoilCol[i,1], ", Horizon - ", specSoilCol[i,2]),
       ylab = "Reflectance",
       xlab = "Wavelength /nm")
}
```

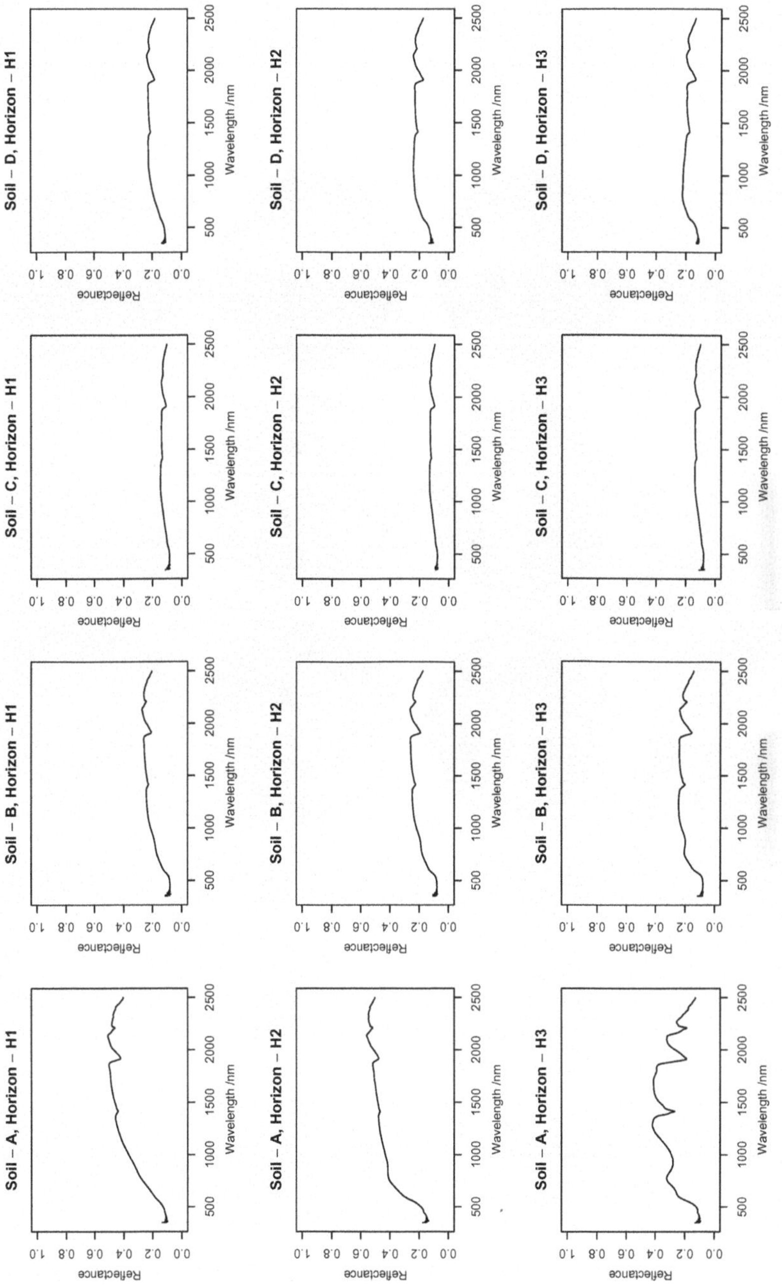

Fig. 6.18 Example reflectance spectra recorded in four different soils and three different depths

Each of these spectra comes from a different soil sample. In total there are 12 samples (four different soils and three depths; see Fig. 6.18). Each spectrum has been recorded by an ASD vis-NIR spectrometer. The spectra provided in this exercise have already been pre-processed, and the replicates have been averaged. The soil colours corresponding to the spectra presented in Fig. 6.18 are presented in Fig. 6.19.

Fig. 6.19 Image of soils used in this example. Left to right by column: Soil A, Brown Sodosol; Soil B, Red Chromosol; Soil C, Black Vertosol; and Soil D, Brown Vertosol. Top to bottom by row: Horizons 1–3

To derive the soil colour from the spectra, we must define the wavelengths representing each colour. The specific wavelength intervals for each colour are provided by Viscarra-Rossel et al. (2009).

```
# provide the intervals for each colour
colourBands <- data.frame(red = c(600,690),
                          green = c(520,600),
                          blue = c(450,520))
```

Now we can plot a spectrum and the associated colour for the intervals. In addition, we add to the colour the mean of each interval. The mean of the interval is useful in the next step (Fig. 6.20).

```
#plot a spectrum so that we can identify the colour bands on the spectrum
matplot(x = colnames(specSoilCol[-c(1:2)]), y = t(specSoilCol[1,-c(1:2)]),
        type = "l",
        ylim = c(0, 1),
        ylab = "Reflectance",
        xlab = "Wavelength /nm")

#loop through each of the 3 RGB colour bands to identify the mean reflectance for each
for(i in 1:ncol(colourBands)){

  #provide the colour code
  cols <- col2rgb(colnames(colourBands[i]))[,1]/255

  # plot the colour band over the spectra
  polygon(c(colourBands[1,i], colourBands[1,i], colourBands[2,i], colourBands[2,i]),
          c(0,1,1,0),
          col = rgb(cols[1], cols[2], cols[3], 0.25))

  # isolate reflectance values within the colour band
  colSpec <- specSoilCol[1, as.character(colourBands[1, i]:colourBands[2, i])]

  # find the mean on the reflectance values
  meanReflectance <- rowMeans(colSpec)

  # plot the mean as a horizontal line
  lines(y = rep(meanReflectance, 2), x = c(colourBands[1,i], colourBands[2,i]),
        col = colnames(colourBands[i]),
        lwd = 3)
}
```

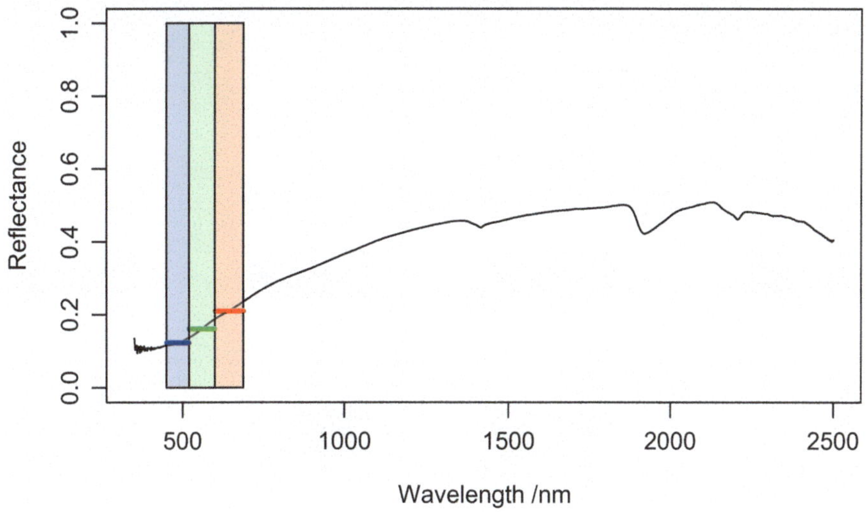

Fig. 6.20 Example reflectance spectra for the soil type A and Horizon H1, with associated blue, green and red colour intervals. The horizontal coloured lines represent the mean of the spectrum within the colour interval

Now that we understand how to extract RGB colours from vis-NIR spectra, we can use these values to estimate colour. We use the function `spectra2colour` from the book package `soilspec` to do this.

```
# return the argument of the spectra2colour function
args(soilspec::spectra2colour)
```

```
## function (spectra, ...)
## NULL
```

As seen above, the function `spectra2colour` takes as argument the reflectance spectra.

```
# the first two columns contain sample ID information that we do not want
# we create a new data frame with those columns removed
colDf <- specSoilCol[,-(1:2)]

# correct for underprediction
# correction factor of 3
colDf <- colDf * 3

# if the correction factor pushed an reflectance values above 1
# needs to be lowered or the function will not run
colDf[colDf > 1] <- 1

# run the spectra2colour() function
cols <- soilspec::spectra2colour(colDf, wavelengths = colnames(colDf))

# set plot parameters to produce an output window
# contains 3 rows, 4 columns and plot consecutively down the column
par(mfcol=c(3,4))

# plot the colours
for(i in 1:nrow(cols)){
  plot(1,
       col = as.character(cols$colour)[i],
       pch = 16,
       cex = 15,
       xlab = NA,
       ylab = NA,
       axes = FALSE)
}
```

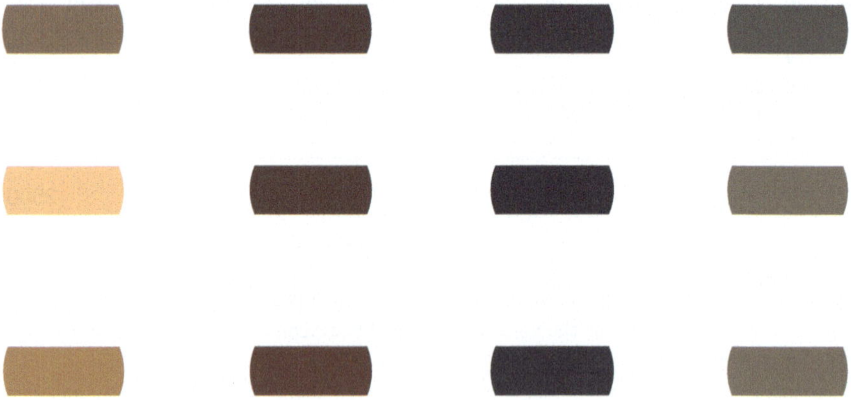

Fig. 6.21 Predicted colours from the visible wavelengths of the spectra. The predicted colours have to be compared with the true colours presented in Fig. 6.19

Figure 6.21 shows a very good representation of the colour of each soil sample. In general, the top soil samples are darker due to increased organic matter content.

References

Abdi H, Williams LJ (2010) Principal component analysis. Wiley Interdiscip Rev Comput Stat 2:433–459

Brown DJ, Shepherd KD, Walsh MG, Mays MD, Reinsch TG (2006) Global soil characterization with VNIR diffuse reflectance spectroscopy. Geoderma 132:273–290

Clark RN, King TVV, Klejwa M, Swayze GA, Vergo N (1990) High spectral resolution reflectance spectroscopy of minerals. J Geophys Res Solid Earth 95:12653–12680

Clark RN, Roush TL (1984) Reflectance spectroscopy: quantitative analysis techniques for remote sensing applications. J Geophys Res Solid Earth 89:6329–6340

Clark RN, Swayze GA, Livo KE, Kokaly RF, Sutley SJ, Dalton JB, McDougal RR, Gent CA (2003) Imaging spectroscopy: earth and planetary remote sensing with the USGS tetracorder and expert systems. J Geophys Res Planets 108:E12

Clark RN, Swayze GA, Wise RA, Livo KE, Hoefen TM, Kokaly RF, Sutley SJ (2007) USGS digital spectral library splib06a. US Geological Survey

Farmer VC (1974) The infrared spectra of minerals. Mineralogical Society of Great Britain, Ireland

Grundy MJ, Viscarra-Rossel RA, Searle RD, Wilson PL, Chen C, Gregory LJ (2015) Soil and landscape grid of Australia. Soil Res 53:835–844

Madejova J, Gates WP, Petit S (2017) Developments in clay science. Elsevier Ltd., pp 107–149. Amsterdam, NL.

Malone BP, Hughes P, McBratney AB, Minasny B (2014) A model for the identification of terrons in the Lower Hunter Valley, Australia. Geoderma Reg 1:31–47

McBratney AB, Mendonça Santos ML, Minasny B (2003) On digital soil mapping. Geoderma 117:3–52

Mouazen AM, Karoui R, Deckers J, De Baerdemaeker J, Ramon H (2007) Potential of visible and near-infrared spectroscopy to derive colour groups utilising the munsell soil colour charts. Biosyst Eng 97:131–143

Ng W, Malone BP, Minasny B (2017) Rapid assessment of petroleum-contaminated soils with infrared spectroscopy. Geoderma 289:150–160

Stenberg B, Viscarra-Rossel RA, Mouazen AM, Wetterlind J (2010) Visible and near infrared spectroscopy in soil science. In: Advances in agronomy. Elsevier, Burlington, pp 163–215

Summers D, Lewis M, Ostendorf B, Chittleborough D (2011) Visible near-infrared reflectance spectroscopy as a predictive indicator of soil properties. Ecol Indic 11:123–131

Varmuza K, Filzmoser P (2016) Introduction to multivariate statistical analysis in chemometrics. CRC Press, Boca Raton

Viscarra-Rossel RA (2011) Fine-resolution multiscale mapping of clay minerals in Australian soils measured with near infrared spectra. J Geophys Res Earth Surf 116:F04023

Viscarra-Rossel RA, Cattle SR, Ortega A, Fouad Y (2009) In situ measurements of soil colour, mineral composition and clay content by vis–NIR spectroscopy. Geoderma 150:253–266

Wold S, Esbensen K, Geladi P (1987) Principal component analysis. Chemom Intell Lab Syst 2:37–52

Chapter 7
Similarity Between Spectra and the Detection of Outliers

In digital soil spectroscopy, similarity or distance metrics between soil spectra are necessary for a large number of applications, such as for assessing the reliability of a spectrometer over repeated scans, to search for a similar soil sample based on spectra from a large database, to classify spectra into groups of similar characteristics or more generally for detecting outlier spectra. To assess similarity, we usually compare two spectra wavelength by wavelength and compute the distance between them. The distance between the two spectra is averaged into a single similarity metric. It is assumed that the closer two spectra are to one to another, the higher is the similarity between the soil properties that they characterize. In the infrared spectroscopy and remote sensing literature, a wide variety of metrics have been deployed. These metrics are based on distances computed directly on the spectra or indirectly from derived information from the spectra.

In this section, we provide an example of distance metrics used to assess the similarity between spectra. For more information on the methods, the reader is redirected to Brereton (2003), to the summary article of Ramirez-Lopez et al. (2013) or to the relevant literature provided in the next pages. Distance metrics are also the basis of several methods described in this chapter (e.g. for outlier detection) or in the next chapters of this book. The metrics described in this section are:

- Euclidean and Mahalanobis distance
- Correlation similarity
- Spectral angle mapper (dot-product cosine distance)
- Spectral information divergence

In addition to describing distance metrics, we show how they are used in combination with principal component analysis (Sect. 6.2) to detect outliers. Distance metrics are useful to determine data points further away from the centre of the spectral principal component space. An outlier is defined as being exceptionally far away from the bulk of the data in the multi-dimensional space of the scores (hence the term multivariate outlier). A multivariate outlier can also be detected as a

A. M. J.-C. Wadoux et al., *Soil Spectral Inference with R*, Progress in Soil Science,
https://doi.org/10.1007/978-3-030-64896-1_7

function of the squared Mahalanobis distance against the chi-square statistic using a multivariate cumulative probability plot. Both procedures, using distance from the centre measures and probability plots, are described using examples.

The set of packages used in this chapter are installed using the lines below. The book-associated `soilspec` package is also required; see Chap. 3 for information on its installation.

```
# specify all the packages used in the chapter and install them if they are not already
myPackages <- c("scatterplot3d", "prospectr", "RcppArmadillo", "resemble",
                "robustbase", "mvoutlier")

#define which packages are not installed in the current computer
notInstalled <- myPackages[!(myPackages %in% installed.packages()[ , "Package"])]

#install the missing packages
if(length(notInstalled)) install.packages(notInstalled)
```

In this section, we use the raw spectra provided by the book-associated `soilspec` package (Fig. 7.1). The steps for pre-processing are explained in the previous chapter.

```
library(soilspec)
data("datsoilspc")

# convert reflectance to absorbance
spectraA <-  log(1/datsoilspc$spc)

# plot first spectrum
matplot(x = colnames(spectraA), y = t(spectraA),
        xlab = "Wavelength /nm",
        ylab = "Absorbance",
        ylim = c(0, 4),
        type = "l",
        lty = 1,
        col = rgb(red = 0.5, green = 0.5, blue = 0.5, alpha = 0.3))
```

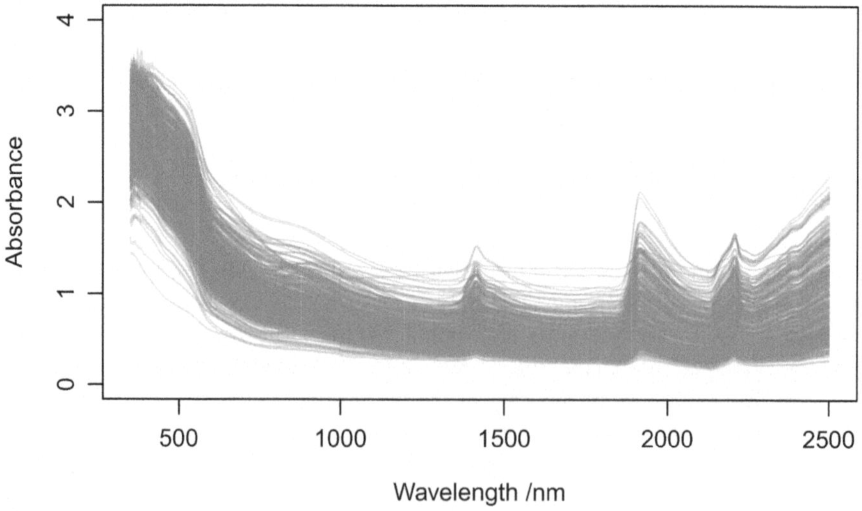

Fig. 7.1 Example set of 391 absorbance soil spectra from the `soilspec` package

Note that in most cases, one would be interested to compute the similarity measures between the principal components of the spectra or between the spectra in a reference dataset and a test set. We provide an example of practical application in the use of distance metrics in Sect. 7.1.6.

7.1 Similarity/Dissimilarity Measures

7.1.1 Euclidean Distance

Euclidean distance is the most common way of representing the distance between two points or objects. It is calculated by taking the square root of the differences between two points. In our case, we compute the pairwise Euclidean distance d between the two spectra $\mathbf{x}_a$ and $\mathbf{x}_b$ with b bands, as follows (Brereton 2003):

$$d_{EU}(\mathbf{x}_a, \mathbf{x}_b) = \sqrt{(\mathbf{x}_b - \mathbf{x}_a)I^{-1}(\mathbf{x}_b - \mathbf{x}_a)^T}, \tag{7.1}$$

where I is the identity matrix. The smaller the distance between the two spectra, the larger the similarity between them. The distance is always positive and has no limit. A small value of the Euclidean distance (i.e. close to zero) indicates strong similarity between the two spectra under study.

The Euclidean distance between spectra can be computed using the `fDiss` function from the `resemble` package. In the following example, we compute the distance between all pairs of spectra contained in the dataset.

```
# load the required package
require(resemble)

# compute Euclidean distance between spectra
EucD <- f_diss(Xr = spectraA,
               Xu = spectraA,
               diss_method = "euclid",
               center = TRUE, scale = TRUE)
```

We can then display the pairwise distance between the first three spectra of the dataset.

```
# print the pairwise distance between the three first spectra
EucD[1:3,1:3]
```

```
##          Xu_1     Xu_2     Xu_3
## Xr_1 0.000000 1.645742 1.352906
## Xr_2 1.645742 0.000000 2.941434
## Xr_3 1.352906 2.941434 0.000000
```

The values of the Euclidean distance displayed above show that the third spectrum of the dataset is more similar to the first spectrum than it is to the second spectrum.

7.1.2 Mahalanobis Distance

Mahalanobis distance (Mahalanobis 1936) is similar to the Euclidean distance but includes a covariance term between spectra, which means that it accounts for the fact that two spectra may measure similar properties due to the correlation between specific bands of the spectra. The Mahalanobis distance between spectra $\mathbf{x}_a$ and $\mathbf{x}_b$ is defined by:

$$d_{MB}(\mathbf{x}_a, \mathbf{x}_b) = \sqrt{(\mathbf{x}_b - \mathbf{x}_a)C^{-1}(\mathbf{x}_b - \mathbf{x}_a)^T}, \tag{7.2}$$

where C is the variance-covariance matrix between the spectra. The variance-covariance matrix is used as a scaling factor of the relationships between two spectra. The use of the covariance matrix introduces additional considerations. In particular, the matrix does not have an inverse when the number of spectra is smaller than the number of wavebands in the dataset. This problem has long been recognized in infrared spectroscopy (see Clark et al. 1993 and Mark and Workman 2010). A simple way around this is to compute the Mahalanobis distance on the first few principal components of the spectra. Principal component analysis is described in Chap. 6.2 and is not repeated here. In this chapter, we use the implementation provided by the function `pc_projection` in the `resemble` package for computing the principal components. The `pc_projection` function requires the user to provide the maximum amount of cumulative variance explained that needs to be retained in the principal components.

```
# specify the maximum amount of variance explained the one want to have retained
maxexplvar <- 0.99
```

Now we can compute the principal components of the absorbance spectra.

```
# compute the principal components
pcspectraA <- pc_projection(Xr = spectraA,
                            pc_selection = list("cumvar", maxexplvar),
                            method = "pca",
                            center = TRUE, scale = FALSE)

# obtain the names of the sub-objects in the PC object created
names(pcspectraA)
```

```
## [1] "scores"        "X_loadings"  "variance"    "scores_sd"    "n_components"
## [6] "pc_selection" "center"      "scale"       "method"
```

The Mahalanobis distance between spectra can be computed by the `fDiss` function, using the PC scores as input.

```
# compute the Mahalanobis distance between spectra scores
mahD <- f_diss(Xr = pcspectraA$scores,
               Xu = pcspectraA$scores,
               diss_method = "mahalanobis",
               center = FALSE, scale = FALSE)
```

We can then display the pairwise distance between the first three scores of the absorbance spectra.

```
# print pairwise distance between the three first spectra
mahD[1:3,1:3]
```

```
##           Xu_1     Xu_2     Xu_3
## Xr_1 0.000000 1.291966 1.074744
## Xr_2 1.291966 0.000000 1.848724
## Xr_3 1.074744 1.848724 0.000000
```

Note that the Mahalanobis distance is always positive. A small value between two spectra indicates that they are similar. For example, the pairwise distance between the first three spectra displayed above shows that the third spectrum of the dataset is more similar to the first spectrum than it is to the second spectrum. The first spectrum is about equally similar to the second and third spectra. This result is similar to that obtained by computing the Euclidean distance, which indicates that the first three spectra are not strongly correlated.

An important note is that the Mahalanobis distance is equivalent to the Euclidean distance computed on the standardized (zero mean and unit variance) principal component scores if the variables are uncorrelated (see Deza and Deza 2009, p. 303). We can test it by specifying `center = TRUE` and `scaled = TRUE` in the `fDiss` function.

```
# compute the Euclidean distance between standardized spectra scores
eucPcD <- f_diss(Xr = pcspectraA$scores,
                 Xu = pcspectraA$scores,
                 diss_method = "euclid",
                 # standardize the PC scores (zero mean and unit variance)
                 center = TRUE, scale = TRUE)

# check that the two distance metrics have same result
all.equal(mahD, eucPcD)
```

```
## [1] TRUE
```

7.1.3 Correlation Similarity

Correlation dissimilarity is based on Pearson's r correlation coefficient between spectra (see also Chap. 9). Pearson's r is not a distance metric, and while two spectra can be strongly correlated, they can have very different reflectance or absorbance values. The value of Pearson's r varies between -1 and 1. A correlation of 1 indicates that the two spectra under study have identical characteristics (i.e. they are similar). A values of -1, conversely, indicates that the two spectra are strongly negatively correlated, which in spectroscopy implies that the two spectra are dissimilar (Brereton 2003). The correlation similarity scales the values between 0 and 1, where the higher the values, the more dissimilar the two spectra are. The correlation dissimilarity (cd) between two spectra can be applied as follows:

$$cd(\mathbf{x}_a, \mathbf{x}_b) = \frac{1 - r(\mathbf{x}_a, \mathbf{x}_b)}{2}. \tag{7.3}$$

This function can be applied manually to the data frame containing the spectra.

```
# compute correlation between spectra
scorrelation <- cor(t(spectraA))

# compute correlation similarity between spectra
cd1 <- (1 - scorrelation)/2
```

```
# print the correlation between three first spectra
round(x = cd1[1:3,1:3], digits = 5)
```

```
##          [,1]    [,2]    [,3]
## [1,] 0.00000 0.04103 0.00368
## [2,] 0.04103 0.00000 0.04700
## [3,] 0.00368 0.04700 0.00000
```

The cd values show that the first spectrum is more similar to the third spectrum than it is to the second spectrum. The largest value is found between the second and third spectra, which means that these two spectra are the most dissimilar between the three first spectra of the dataset.

The same results is obtained by the `resemble` package with the `corDiss` function.

```
# compute correlation similarity with the resemble package
cd2 <- cor_diss(Xr = spectraA,
                Xu = spectraA,
                center = FALSE, scale = FALSE)

# check that the two methods have similar results
round(x = cd2[1:3,1:3], digits = 5)
```

```
##         Xu_1    Xu_2    Xu_3
## Xr_1 0.00000 0.04103 0.00368
## Xr_2 0.04103 0.00000 0.04700
## Xr_3 0.00368 0.04700 0.00000
```

Alternatively, the correlation similarity can be computed using a moving window. In this case, the correlation similarity is computed by averaging the moving window correlation measures. It can be computationally demanding for large datasets. We start by defining a window size.

```
# the moving window value must be an odd number
w4cd <- 51
```

Now we can use it in the corDiss function in resemble.

```
# compute the correlation similarity with a moving window
mwcd <- cor_diss(Xr = spectraA,
                 Xu = spectraA,
                 ws = w4cd,
                 center = FALSE, scale = FALSE)

# check the first three correlation similarity values
mwcd[1:3,1:3]
```

```
##             Xu_1       Xu_2       Xu_3
## Xr_1 0.0000000 0.18195370 0.07498450
## Xr_2 0.1819537 0.00000000 0.09979849
## Xr_3 0.0749845 0.09979849 0.00000000
```

Using the moving window to compute the correlation similarity measure provides much different values than when computed on the whole spectrum. While the values are different, the first spectrum is still more similar to the second spectrum than it is to the third spectrum. The second spectrum seems not to be strongly correlated to the third spectrum. This is different from the values obtained using the correlation similarity measure computed on the whole spectra.

7.1.4 Spectral Angle Mapper

The spectral angle mapper (SAM) also called dot-product cosine distance (Yuhas et al. 1992; Farifteh et al. 2007) has been extensively applied in remote sensing as a tool for unsupervised classification and spectral similarity analysis. The SAM algorithm derives an angular difference (in radians) between two spectra, which means that the larger the angle is, the more dissimilar the two spectra are. The SAM distance between two spectra ($\mathbf{x}_a$ and $\mathbf{x}_b$) is calculated as:

$$\mathrm{SAM}(\mathbf{x}_a, \mathbf{x}_b) = \cos^{-1} \frac{\sum_{j=1}^{b} \mathbf{x}_{a,j}\mathbf{x}_{b,j}}{\sum_{j=1}^{b}(\mathbf{x}_{a,j}^2)^{1/2} \sum_{j=1}^{b}(\mathbf{x}_{a,j}^2)^{1/2}}. \quad (7.4)$$

where $j = 1, ..., b$ is the number of spectral bands. The values range between 0 and $\pi/2$ (about 1.57), where small values indicates strong similarity. The range of values can be scaled between 0 and 1 by dividing $\pi/2$ to π (Schwarz and Staenz 2001).

This is implemented in the fDiss function in the resemble package.

```
# compute the pairwise SAM similarity
samD <- f_diss(Xr = spectraA,
               Xu = spectraA,
               diss_method = "cosine",
               center = FALSE, scale = FALSE)

# show the first three similarity values
samD[1:3,1:3]
```

```
##             Xu_1      Xu_2      Xu_3
## Xr_1 0.0000000 0.3293496 0.1252271
## Xr_2 0.3293496 0.0000000 0.4131082
## Xr_3 0.1252271 0.4131082 0.0000000
```

The values displayed above show that the first spectrum is more similar to the third spectrum than it is to the second one. The largest dissimilarity between the three first spectra is obtained between the second and third spectra.

7.1.5 Spectral Information Divergence

Spectral information divergence (SID) was introduced by Chang (2000) to measure the similarity between spectra based on their spectral signature probability distribution. In other words, it measures the distance between the probability distribution produced by the spectral signature of two vectors (Chang 2003). To measure the distance, the Kullback-Leibler (KL) divergence (Kullback and Leibler 1951) or cross-entropy is used. The SID is defined by (Chang 2003):

$$\mathrm{SID}(\mathbf{x}_a, \mathbf{x}_b) = \mathrm{KL}(\mathbf{x}_a||\mathbf{x}_b) + \mathrm{KL}(\mathbf{x}_b||\mathbf{x}_a) \tag{7.5}$$

$$= \sum_{j=1}^{b} p_j \log \frac{p_j}{q_j} + \sum_{j=1}^{b} q_j \log \frac{q_j}{p_j} \tag{7.6}$$

where b is the total number of bands and $\mathbf{p} = \mathbf{x}_a / \sum_{j=1}^{b} \mathbf{x}_{a,j}$ and $\mathbf{q} = \mathbf{x}_b / \sum_{j=1}^{b} \mathbf{x}_{b,j}$ are the probability vectors (vector with non-negative entries that add up to one) of $\mathbf{x}_a$ and $\mathbf{x}_b$, respectively. $\mathrm{KL}(\mathbf{x}_a||\mathbf{x}_b)$ is the Kullback-Leibler (1951) information measure. The higher the value, the larger is the difference (the dissimilarity) between the two spectra.

This is implemented in the sid function in the resemble package.

```
# compute the pairwise SID similarity
sidD <- sid(Xr = spectraA,
            Xu = spectraA,
            mode = "density",
            center = FALSE, scale = TRUE)

# show the first three similarity values
sidD$sid[1:3,1:3]
```

```
##          Xu_1     Xu_2     Xu_3
## Xr_1 0.000000 5.525884 5.707669
## Xr_2 5.525884 0.000000 5.844254
## Xr_3 5.707669 5.844254 0.000000
```

The values displayed above show that the first spectrum is more similar to the second spectrum than it is to the third one. The largest dissimilarity between the three first spectra is obtained between the second and third spectra. Note that by specifying `mode = "density"`, the Kullback-Leibler information measure is computed on the probability density of the spectra (default), while by specifying `mode = "feature"`, the Kullback-Leibler information measure is computed directly on the spectra.

Several other similarity or distance measures can evaluate the discrepancy between spectra. In particular, a combination of the SAM and SID (referred to as a SAM-SID measure) has been developed by Du et al. (2004):

$$\mathrm{SAMSID}_{\tan}(\mathbf{x}_a, \mathbf{x}_b) = \mathrm{SID}(\mathbf{x}_a, \mathbf{x}_b)\tan(\mathrm{SAM}(\mathbf{x}_a, \mathbf{x}_b)) \tag{7.7}$$

```
samsidTAN <- sidD$sid*tan(samD)
samsidTAN[1:3, 1:3]
```

```
##           Xu_1     Xu_2      Xu_3
## Xr_1 0.0000000 1.888738 0.7185146
## Xr_2 1.8887382 0.000000 2.5617209
## Xr_3 0.7185146 2.561721 0.0000000
```

or:

$$\mathrm{SAMSID}_{\sin}(\mathbf{x}_a, \mathbf{x}_b) = \mathrm{SID}(\mathbf{x}_a, \mathbf{x}_b)\sin(\mathrm{SAM}(\mathbf{x}_a, \mathbf{x}_b)) \tag{7.8}$$

```
samsidSIN <- sidD$sid*sin(samD)
samsidSIN[1:3, 1:3]
```

```
##           Xu_1     Xu_2      Xu_3
## Xr_1 0.0000000 1.787224 0.7128881
## Xr_2 1.7872240 0.000000 2.3462224
## Xr_3 0.7128881 2.346222 0.0000000
```

The values of $SAMSID_{tan}$ or $SAMSID_{sin}$ displayed above show that the first spectrum is more similar to the third spectrum than it is to the second one. The strongest dissimilarity is found between the second and third spectra. The product of SAM and SID is supposed to increase the spectral discriminability. Small values of $SAMSID_{tan}$ or $SAMSID_{sin}$ indicate similarity between the spectra. More detailed description of the measures is provided in the Chang (2003) textbook.

7.1.6 A Practical Example

We provide an example on the use of the similarity/dissimilarity measures similar to the one in Ramirez-Lopez et al. (2013) using the `resemble` package. Ramirez-Lopez et al. (2013) argued that the similarity measures between the spectra should be able to reflect their similarity also in terms of ancillary information, for example, the soil properties associated with the spectra. To test this, the authors proposed (Fig. 3 in Ramirez-Lopez et al. (2013)) to:

1. Compute the principal components for the whole set of spectra.
2. Compute a distance matrix based on the principal components.
3. Split randomly the dataset into two subsets. The first subset is called `xuData` and contains about 25% of the original spectra and associated measured soil property. The second subset is called `xrData` and contains 75% of the spectra and the soil properties.
4. For each spectrum in `xuData`, select the closest spectrum in `xrData`.
5. Compare the value of the soil property from subset `xuData` to the values of the soil property for the closest spectra selected at step 4.

For this example, we use the reflectance spectra and associated values of the total carbon contained in the `soilspec` package. The dataset information can be accessed by `?soilspec::datsoilspc`.

We start by loading the data.

```
# load the required package
require(soispec)

# take the organic carbon values and the associated spectra (stored in a matrix).
specACarbon <- datsoilspc[,c("TotalCarbon", "spc")]

# create a vector containing the wavelength
wavs <- as.numeric(colnames(specACarbon$spc))
```

To test the approach, we start by dividing the dataset into two disjoint subsets. They are divided by random selection of their row number.

```
# take a number of rows (i.e., number of spectra)
ns <- nrow(specACarbon)

# indicate the percentage that you want to select
pd <- 0.25
```

```
# compute the integer number corresponding to the 25% of the samples
nsamples2select <- round(x = ns * pd, digits = 0)

# compute the vector of sample indexes
origIndexes <- 1:ns

# randomly select the samples
set.seed(19)
selSampleIndx <- sample(x = origIndexes, size = nsamples2select)

# first the indexes randomly found
xuData <- specACarbon[selSampleIndx,]

# then the rest (i.e. the ones that were not selected)
xrData <- specACarbon[-selSampleIndx,]
```

To compute the Mahalanobis distance between the spectra, we first need to compute their principal components. Since we only need the spectra to compute the principal components, we can do it on the combined `xuData` and `xrData` data.

```
# combine the spectral data of xrData  and xuData (by rows) into one single dataset
combX <- rbind(xrData$spc, xuData$spc)

# compute the PCs of the combined spectra
maxexplvar <- 0.999

# compute the principal components
pcspectra <- pc_projection(Xr = combX,
                           pc_selection = list("cumvar", maxexplvar),
                           method = "pca",
                           center = TRUE, scale = FALSE)

# obtain the names of the sub-objects in the PC object created
names(pcspectra)
```

```
## [1] "scores"       "X_loadings" "variance"     "scores_sd"    "n_components"
## [6] "pc_selection" "center"     "scale"        "method"
```

Note here that the first 293 rows of the score matrix correspond to the scores of the spectra of `xrData` and the remaining ones correspond to the scores of the spectra for `xuData`.

We can now plot the two first scores of each dataset.

```
plot(x = pcspectra$scores[1:nrow(xrData), 1],
     y = pcspectra$scores[1:nrow(xrData), 2],
     xlab = "PC 1",
     ylab = "PC 2",
     type = "p",
     pch = 16,
     col = rgb(red = 0, green = 0.4, blue = 0.8, alpha = 0.5))
grid()

# add to the plot the first two scores of the PCs of the xuData
points(x = pcspectra$scores[-c(1:nrow(xrData)), 1],
       y = pcspectra$scores[-c(1:nrow(xrData)),2],
       xlab = "PC 1",
       ylab = "PC 2",
       pch = 16,
       col = rgb(red = 0.8, green = 0.4, blue = 0, alpha = 0.5))
```

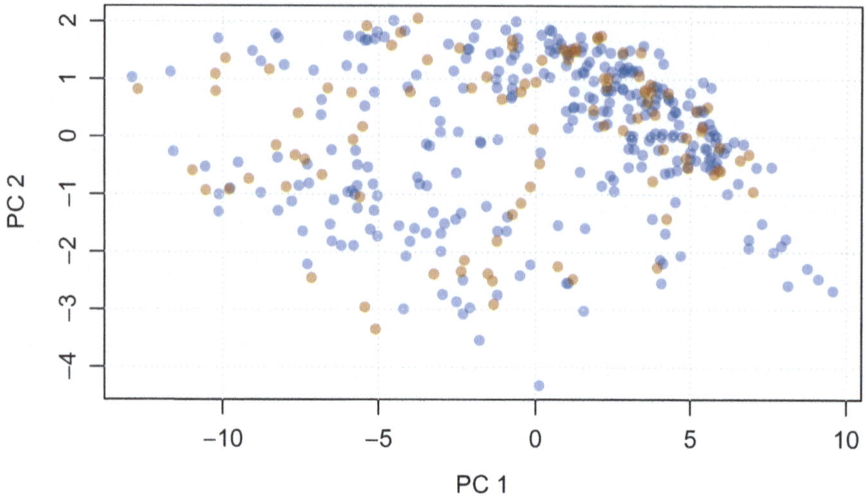

Fig. 7.2 First two principal components of the absorbance spectra. The blue dots correspond to the spectra from `xrData` and the orange dots to the spectra from `xuData`

Now that we have the principal components, we can compute the Mahalanobis pairwise distance matrix for the spectral data.

```
# compute the  Mahalanobis pairwise distance from the xrData dataset
mdXr <- f_diss(Xr =  pcspectra$scores,
               Xu =  pcspectra$scores,
               diss_method = "mahalanobis",
               center = FALSE, scale = FALSE)
```

Note here that `mdXr` is the dissimilarity matrix of the spectra computed in its PC space. Now we can select for each spectrum in `xuData` its closest (the most spectrally similar) spectra in `xrData` using the dissimilarity measures stored in `mdXr`.

```
# first create an empty object to store the results of the indices of nearest neighbours
nearestN <- NULL

# use a for loop to iterate over each of the columns in the mdXr matrix
# use only the rows of xuData
for(i in (nrow(xrData)+1):nrow(mdXr)){

  # order the values from smallest distance to the largest
  # take the second one (first value is the spectrum itself)
  nn_i <- order(x = mdXr[1:nrow(xrData),][,i])[2]

  # store the results in the nearestN object
  nearestN <- c(nearestN, nn_i)
}
```

The resulting `nearestN` object is a vector containing the indices of the nearest neighbours for each of the spectra in `xrData`.

```
# show the ten most similar spectra
nearestN[1:10]
```

```
##  [1]   7 241 274 190 205 156 193  24 264 155
```

We can now compare the soil property value (i.e. total carbon) of each spectrum to the soil property value for the associated closest sample. If the original dataset is sufficiently large and the similarity measures are reliable, the soil property values from both sources should be similar.

We first want to obtain the soil property value of the nearest neighbours found in `xrData` for each spectrum in `xrData` by `xrData[nearestN,"Organic carbon"]`. We can plot the soil property values contained in `xrData` and in the associated closest neighbour (Fig. 7.2).

```
plot(x = xuData[,"TotalCarbon"],
     y = xrData[nearestN,"TotalCarbon"],
     xlab = "Total carbon reference spectra, /%",
     ylab = "Total carbon from nearest neighbour, /%",
     pch = 16,
     col = rgb(red = 1, green = 0.2, blue = 0.2, alpha = 0.5))
grid()
```

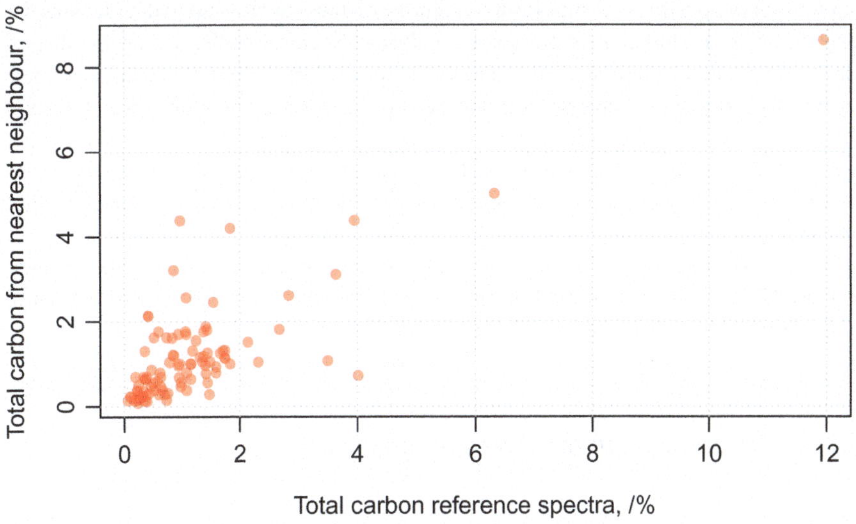

Fig. 7.3 Scatterplot of the values of the carbon content from `xrData` against the values of carbon content found from its closest spectra in the `xuData` dataset

The `eval` function from the `soilspec` package can be used to assess the reliability of the dissimilarity measures by computing a set of accuracy measures.

```
require(soilspec)

# use the eval function to get a set of accuracy measures
dEval <- eval(obs = xrData[nearestN,"TotalCarbon"],
              pred = xuData$"TotalCarbon", obj = "quant")

# print the results
dEval
```

```
##      ME RMSE   r2   R2 rhoC  RPD RPIQ
## 1  0.01 0.91 0.44 0.46 0.77 1.36 1.19
```

We can see that there is a strong correlation, as assessed by the square of Pearson's r correlation coefficient (r^2). This means that the Mahalanobis distance is efficient in detecting neighbouring spectra. The selected neighbouring spectra carry similar information as shown in Fig. 7.3.

7.2 Detecting Outlier Spectra

An outlier spectrum is a spectrum that is significantly distant (dissimilar) from the population of spectra. It is important to identify distant spectra and to decide whether they must be included in subsequent analyses. Spectral outliers may be due to variations related to sample heterogeneity, to spectral measurement error (e.g. noise in the spectrometer, improper optical set-up) or to chemical/physical variation in the sample (e.g. the soil sample has been taken by mistake from a different area). We stress here that it is important to define the cause of the dissimilarity of the spectra before taking any decision on whether removing it from further analysis.

In this section, we focus only on multivariate outliers, i.e. we do not present univariate outlier detection methods (e.g. using the inter-quartile range of the data), which are of less relevance in digital soil spectroscopy.

7.2.1 *Distance from the Average Spectrum*

To detect outliers, we first show how to use the distance from the centre of the multi-dimensional space as an estimate of the dissimilarity between spectra. If a spectrum is far away from the average spectrum, it is a potential outlier. In spectroscopy outlier detection is usually applied in the principal component space. The distances are then computed in the multi-dimensional space of the principal component scores with respect to the underlying covariance matrix. The origin of this multi-dimensional space is the mean of the principal component scores.

We start by applying principal component analysis before computing the distances. We will apply the Mahalanobis distance and the *H* distance to identify a spectrum as an outlier. Note that the Mahalanobis distance is based on the arithmetic mean and sample covariance structure, which are both sensitive to outliers. In most cases, the distance should be robust, i.e. computed using a robust estimate of the mean and covariance structure. We provide an example later in this chapter.

We use the absorbance spectra from the previous example (`soilspec` package) and apply the standard normal variate pre-processing (Sect. 5.2).

```
# load the require package
require(prospectr)

# apply a standard normal variate transformation
spectraASnv <- standardNormalVariate(spectraA)
```

We can compute the principal components of the spectra as before, choosing the minimum amount of variance that we want to retain. In our case, we have chosen the amount of variance to retain so that we obtain three principal components. We made this choice for illustrative purposes.

```
# choose the amount of variance explained
maxexplvar <- 0.99

# compute the principal components
pcspectraA <- pc_projection(Xr = spectraASnv,
                            pcSelection = list("cumvar", maxexplvar),
                            center = TRUE, scale = FALSE)
pcspectraA
```

```
##
##  Method:  pca (svd)
##  Number of components retained:  6
##  Number of observations and number of original variables:  391 2151
##
##  Standard deviations, cumulative variance explained, individual variance explained:
##
##  Explained variance in X {Xr; Xu}:
##                            pc_1  pc_2   pc_3   pc_4    pc_5    pc_6
## sd                        7.128 3.192 1.7067 1.0180 0.75933 0.50000
## cumulative_explained_var 0.763 0.916 0.9602 0.9758 0.98446 0.98822
## explained_var            0.763 0.153 0.0438 0.0156 0.00866 0.00376
```

In the next step, we want to derive the mean of each PC scores. The mean of each PC score will be the average of the spectra, our reference for the outlier detection.

```
# calculate the average of the PC scores
pcspectraACentre <- colMeans(pcspectraA$scores)

# since the result of the 'colMeans' function is a vector
# reformat it to a matrix of 1 row
pcspectraACentre <- t(as.matrix(pcspectraACentre))

# print the results
print(pcspectraACentre)
```

```
##             pc_1          pc_2         pc_3          pc_4          pc_5
## [1,] 2.168804e-16 -2.387796e-16 1.627357e-17 -1.770217e-16 -1.347062e-16
##              pc_6
## [1,] -1.099021e-16
```

Three components retain 96% of the variance in this set of infrared spectra. The centre of the three PC scores is now used to calculate the dissimilarity between each spectrum and the average values of the spectra. All this is done in the principal component space of the spectra.

The Mahalanobis Distance

The description of the Mahalanobis distance is provided in Sect. 7.1.2. We start by computing the Mahalanobis distance between the centre and all principal component scores of the spectra.

```
# compute Mahalanobis distance between scores centre and the scores of the spectra
wmahald <- f_diss(Xr =  pcspectraA$scores,
                  Xu =  pcspectraACentre,
                  diss_method = "mahalanobis",
                  center = FALSE, scale = FALSE)
```

The critical step is to decide whether a spectrum can be considered as an outlier. Most methods use an arbitrary distance beyond which a spectrum is considered an outlier. Again, the major reason to consider a spectrum as an outlier should not be based on an arbitrary limit but on critical reasoning.

We generally consider spectra with a Mahalanobis distance larger than 3 as an outlier. We can plot the dissimilarity scores vs the index of the sample (Fig. 7.4).

```
# plot the index of the spectra against the Mahalanobis distance
plot(wmahald,
     pch = 16,
     col = rgb(red = 0, green = 0.4, blue = 0.8, alpha = 0.5),
     ylab = "Mahalanobis distance")

# add a horizontal line
# visualize the spectra with Mahalanobis dissimilarity scores larger than 3
# (arbitrary threshold)
abline(h = 3, col = "red")
```

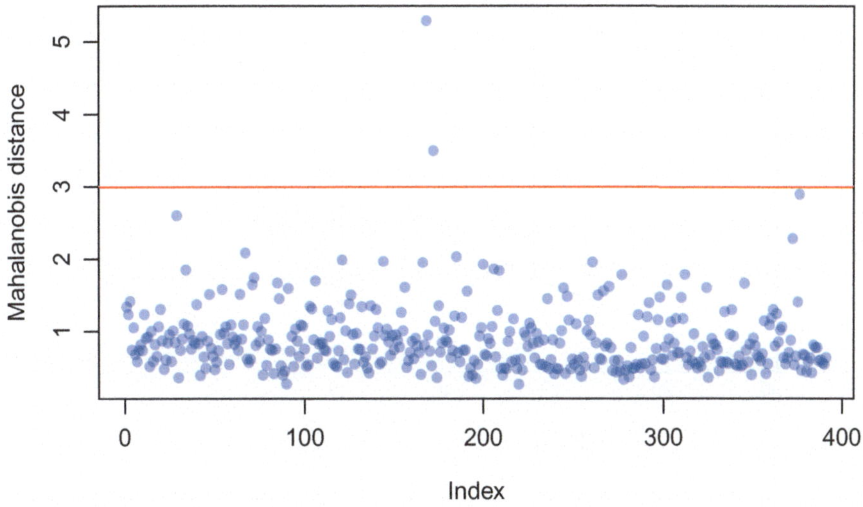

Fig. 7.4 Index of the spectra against the Mahalanobis distance between the spectra principal component scores and the centre of the scores. The points above the horizontal red line have a Mahalanobis distance from the mean of the PC scores greater than 3

We can identify the number of samples with a value of the Mahalanobis distance larger than 3.

```
# obtain the indices of the outliers
indxOutM <- which(wmahald > 3)

# how many potential outliers?
length(indxOutM)
```

```
## [1] 2
```

To visualize if this outlier detection limits is realistic, we can plot the potential outliers in the PC space of the three principal component scores (Fig. 7.5).

```
# load the required package
require(scatterplot3d)

# plot the first three PCs along the identified outliers
sct3d <-scatterplot3d(pcspectraA$scores[,1],
                      pcspectraA$scores[,2],
                      pcspectraA$scores[,3],
                      xlab = "PC 1",
                      ylab = "PC 2",
                      zlab = "PC 3",
                      color=rgb(red = 0.5, green = 0.5, blue = 0.5, alpha = 0.5),
                      pch = 16, grid=TRUE, angle=50)

# add location of the outliers in red
sct3d$points3d(pcspectraA$scores[indxOutM,1],
               pcspectraA$scores[indxOutM,2],
               pcspectraA$scores[indxOutM,3],
               pch = "X",
               col = "red")
```

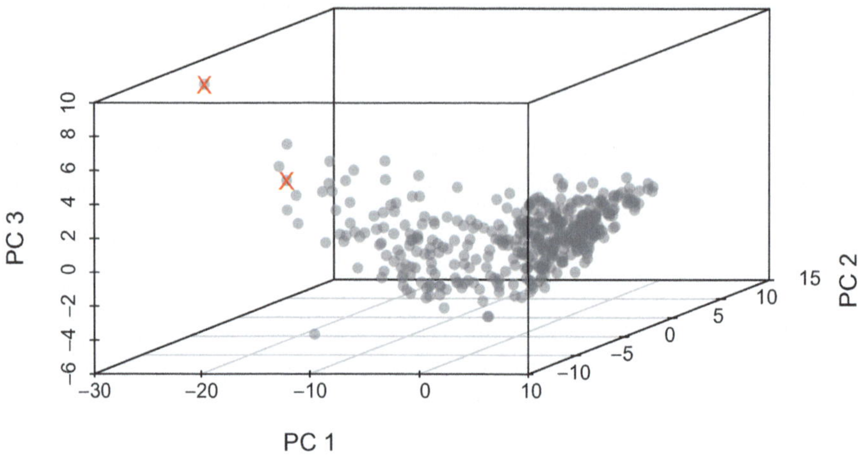

Fig. 7.5 Location of the potential outliers from the three first principal component score spaces of the absorbance spectra

The selected outliers have a red cross. The selected outliers seem to be far away from the bulk of the data in the principal component space.

H Distance

Similar to the Mahalanobis distance, the *H* distance (Mark 1986; Shenk and Westerhaus 1991) can be used to detect outliers. The *H* distance is the square root of the product between the squared Mahalanobis distance and the number of principal components.

```
# compute the H distance
hs <- (wmahald^2 * pcspectraA$n_components)^0.5
```

Like the arbitrary number of 3 for detecting outlier using the Mahalanobis distance, we use a number of 6 to detect possible outliers using the *H* distance. We can plot the *H* score value and the index of the spectra as before (Fig. 7.6).

```
# plot the index of the spectra against the H distance
plot(hs,
     pch = 16,
     col = rgb(red = 0.5, green = 0.5, blue = 0.5, alpha = 0.5),
     ylab = "H distance")

# add a horizontal line
# visualize the spectra with H distance larger than 6 (arbitrary threshold)
abline(h = 6, col = "red")
```

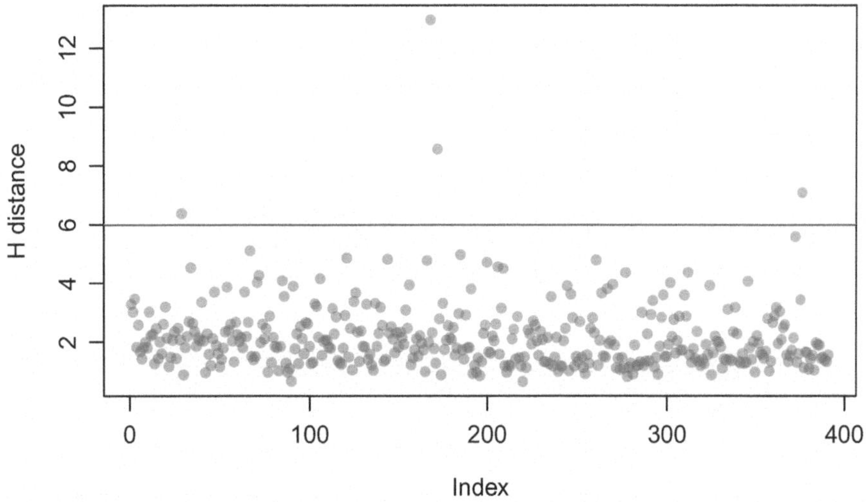

Fig. 7.6 Index of the spectra against the *H* distance. The points above the horizontal red line have a *H* distance greater than 6

We can identify how many spectra have a value of the *H* distance larger than 3 (Mark 1986).

```
# obtain the indices of the outliers
indxOutH <- which(hs > 6)

# how many potential outliers?
length(indxOutH)
```

```
## [1] 4
```

We can plot the potential outliers in the PC scores space (Fig. 7.7).

```
# load the required package
require(scatterplot3d)

# plot the first three PCs along the identified outliers
sct3d <-scatterplot3d(pcspectraA$scores[,1],
                      pcspectraA$scores[,2],
                      pcspectraA$scores[,3],
                      xlab = "PC 1",
                      ylab = "PC 2",
                      zlab = "PC 3",
                      color=rgb(red = 0.5, green = 0.5, blue = 0.5, alpha = 0.5),
                      pch = 16, grid=TRUE, angle=50)

sct3d$points3d(pcspectraA$scores[indxOutH,1],
               pcspectraA$scores[indxOutH,2],
               pcspectraA$scores[indxOutH,3],
               pch = "X",
               col = "red")
```

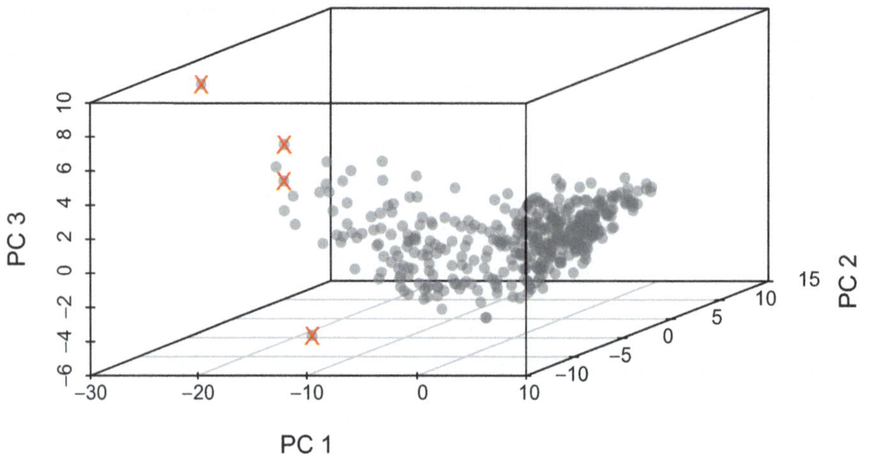

Fig. 7.7 Location of the potential outliers from the three first principal component score spaces of the absorbance spectra

Using the H distance, it seems that similar points are selected to those chosen with the Mahalanobis distance. It also appears that one point is close to the arbitrary limit of 6. The visualization shows that this spectrum is probably not an outlier as it is well contained in the bulk of the data.

7.2.2 *Multivariate Outliers*

The Mahalanobis distance methods have some drawbacks. In particular, the user must choose an arbitrary cutoff value above which a spectrum is considered an outlier. These simple methods are based on the distance in the multivariate space, but do not distinguish between extremes of a distribution and outliers, which potentially come from a different distribution (Rousseeuw and Driessen 1999).

To detect multivariate outliers, it is possible to assume that the multivariate distribution of the principal component scores follows a multivariate normal distribution with a given mean and covariance matrix. Garrett (1989) showed that the squared Mahalanobis distance follows a χ^2_m distribution with m degrees of freedom, where m is the number of principal components. Values of the squared Mahalanobis distance higher than the 97.5% quantile of the χ^2_m distribution are considered exceptionally high and likely to be outliers. This method is largely interactive as the user must define manually what makes a good cutoff value (if not using the 97.5% quantile). Similar adaptive methods which account for different sample sizes have also been proposed, and they are discussed later in this section.

With the scripts below, this is implemented first by calculation of the Mahalanobis distance between the spectra using the Mahalanobis function from the base `stats` package. Note that this is different from the previous methods where the distances are computed between the centre and all the principal component scores. Both the arithmetic mean and sample covariances are sensitive to outliers in the data. Garrett (1989) proposed using a robust estimate of the mean and sample covariance matrix of the principal component scores. Ideally, we can also use a robust computation of the principal components. How to compute robust principal components is not covered in this book, and the reader is redirected to Varmuza and Filzmoser (2016), Section 3.6, for an introduction and the associated R codes.

```
# load the required packages
library(robustbase)

# how many components are we using?
ncomp <- pcspectraA$n_components

# calculate the average of the PC scores
pcspectraACentre <- covMcd(pcspectraA$scores)$center
pcspectraACentre
```

```
##        pc_1        pc_2        pc_3        pc_4        pc_5        pc_6
##  4.52142968 -0.32311440  0.14471836 -0.10692539  0.06974600  0.05176857
```

```
# derive the covariance between of the principal component scores
pcspectraACov <- covMcd(pcspectraA$scores)$cov
pcspectraACov
```

```
##             pc_1        pc_2       pc_3         pc_4       pc_5        pc_6
## pc_1 3.829571811  2.25153074  1.4546086  0.004683711  0.0747860  0.35268376
## pc_2 2.251530742  6.70624805 -0.4502221 -0.286918035  0.2241723  0.06238531
## pc_3 1.454608568 -0.45022213  3.0697179  0.172406599  0.1677416 -0.41200932
## pc_4 0.004683711 -0.28691803  0.1724066  0.444332846  0.2649986 -0.07467132
## pc_5 0.074785997  0.22417227  0.1677416  0.264998554  0.2821209 -0.11547914
## pc_6 0.352683758  0.06238531 -0.4120093 -0.074671319 -0.1154791  0.21651079
```

```
# create an empty matrix
chiMat <- matrix(NA, ncol = 2, nrow = nrow(spectraA))

# compute squared Mahalanobis distance
chiMat[,1] <- sqrt(mahalanobis(pcspectraA$scores,
                               pcspectraACentre,
                               pcspectraACov))

# sort the distance
chiMat[,1] <- sort(chiMat[,1])
```

Then we derive the cumulative χ^2 probability function of the Mahalanobis distances. Here we use the base `pchisq` function. This is followed by estimation of the `cutoff` value for specifying which spectra are and are not possible outliers. The value in `cutoff` corresponds to a value of $\chi^2_{m=6,0.975} = 14.45$

```
# chi squared distribution with ncomp degrees of freedom (df = ncomp)
chiMat[,2] <- pchisq(chiMat[,1], df =  ncomp)
```

```
# take the 0.975 quantile of the chi square distribution with m = 5
cutoff <- qchisq(0.975, ncomp)

# which spectra are outliers
outliers <- which(chiMat[,1] >= cutoff)

# how many outliers?
length(outliers)
```

```
## [1] 97
```

This method detects a large number of outliers. Plotting the cumulative χ^2 function against Mahalanobis distance gives the plot below (Fig. 7.8). The points marked with the red crosses are the outliers, and the green line is the cutoff distance.

```
# plot the cumulative chi^2 function against the Mahalanobis distance
plot(chiMat[,1],
     chiMat[,2],
     xlab = "Mahalanobis distance",
     ylab = "Cumulative probability",
     col = rgb(red = 0.5, green = 0.5, blue = 0.5, alpha = 0.5),
     pch = 16)

# add cross to the points that can be considered as outliers
points(chiMat[outliers,],
       pch="X",
       col="red")

# add the 0.975 cutoff limit
abline(v = cutoff,
       col = "#3333CC")
 text(x = cutoff, y = 0.4, "97.5% quantile", col = "#3333CC",
      pos = 4, srt = 90, cex = 0.8)
```

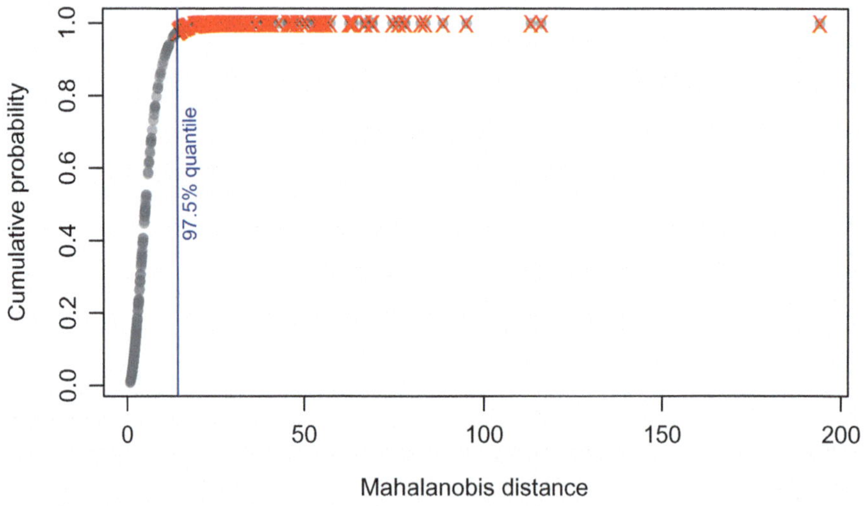

Fig. 7.8 Plot of the cumulative χ^2_m function against Mahalanobis distance. The red crosses indicate possible outliers, and the blue line is the 97.5% quantile cutoff distance

The plot below is the three-dimensional plot of the principal component scores where the points marked with the red cross are the outliers. Intuitively however, without performing this outlier removal procedure, we could probably ascertain most of these outlier points (Fig. 7.9).

```
# load the required package
require(scatterplot3d)

# plot the first three PCs along the identified outliers
sct3d <-scatterplot3d(pcspectraA$scores[,1],
                      pcspectraA$scores[,2],
                      pcspectraA$scores[,3],
                      xlab = "PC 1",
                      ylab = "PC 2",
                      zlab = "PC 3",
                      color= rgb(red = 0.5, green = 0.5, blue = 0.5, alpha = 0.5),
                      pch = 16, grid=TRUE, angle=50)
# add red cross to potential outliers
sct3d$points3d(pcspectraA$scores[,1][outliers],
               pcspectraA$scores[,2][outliers],
               pcspectraA$scores[,3][outliers],
               pch = "X",
               col = "red")
```

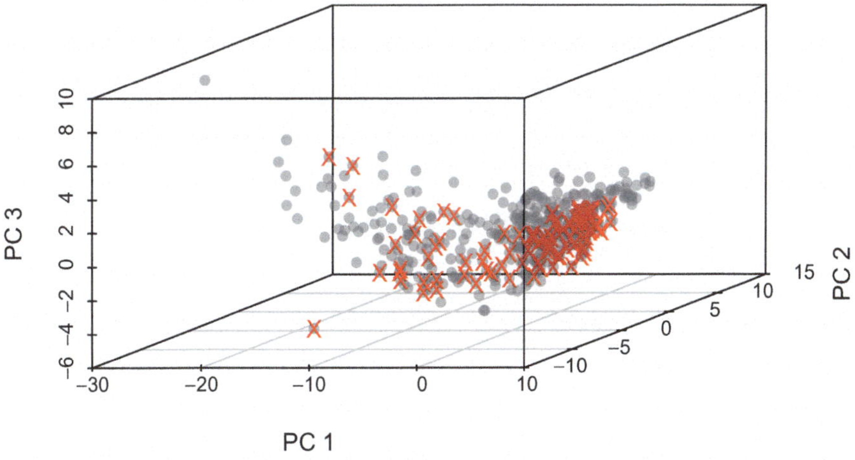

Fig. 7.9 Location of the potential outliers from the three first principal component score spaces of the absorbance spectra

The example based on the approach by Garrett (1989) suffers from several drawbacks. In particular, the cutoff value is arbitrary and must be adjusted to the dataset. The method also does not adjust for the sample size. To solve this, an adaptive method was proposed by Filzmoser et al. (2005) where the cutoff value is defined by a measure of deviation of the empirical distribution function of Mahalanobis distance from the theoretical distribution function, defined by Filzmoser et al. (2005) and implemented in the `mvoutlier` package by the `arw` function.

```
# load the required package
require(mvoutlier)

# compute the distance for the threshold
distCutoff <- mvoutlier::arw(pcspectraA$scores,
                             m0 = pcspectraACentre,
                             c0 = pcspectraACov,
                             alpha = 0.05)$cn
distCutoff
```

```
## [1] 17.77791
```

From the distance, we can obtain the cutoff value from the χ^2 distribution.

```
# compute the cutoff value at the threshold distance
AdjCutoff <- pchisq(distCutoff, df = ncomp)
AdjCutoff
```

```
## [1] 0.9931882
```

We see that the value of the cutoff is more conservative than the usual or non-adaptive 0.975 cutoff, which means that some values were previously considered as outliers are not considered so; they are only extremes of the distribution.

Now that we know both the threshold distance and cutoff value of the χ^2 distribution, we can do as before and find the outliers.

```
# which spectra are outliers
outliers <- which(chiMat[,1] >= distCutoff)

# how many outliers?
length(outliers)
```

```
## [1] 87
```

Less spectra are considered outliers. This number is still considerable, and one should try to understand why many spectra are considered as outliers. Perhaps using a robust estimate of the principal components or by applying some smoothing and other pre-processing to the spectra would markedly reduce this number.

We can plot the Mahalanobis distance against the empirical distribution function of the distance. We add both the adjusted cutoff value and the original 97.5% quantile value for illustration (Fig. 7.10).

```
# plot the cumulative chi^2 function against the Mahalanobis distance
plot(chiMat[,1],
     chiMat[,2],
     xlab = "Mahalanobis distance",
     ylab = "Cumulative probability",
     col = rgb(red = 0.5, green = 0.5, blue = 0.5, alpha = 0.5),
```

```
      pch = 16)

# add cross to the points that can be considered as outliers
points(chiMat[outliers,],
       pch="X",
       col="red")

# add the adjusted cutoff limit
abline(v = distCutoff,
       col ="#006600")
text(x = distCutoff, y = 0.4, "Adjusted Quantile", col = "#006600",
      pos = 4, srt = 90, cex = 0.8)

# add the 0.975 cutoff limit
abline(v = cutoff,
       col ="#3333CC")
text(x = cutoff, y = 0.4, "97.5% quantile", col = "#3333CC",
      pos = 4, srt = 90, cex = 0.8)
```

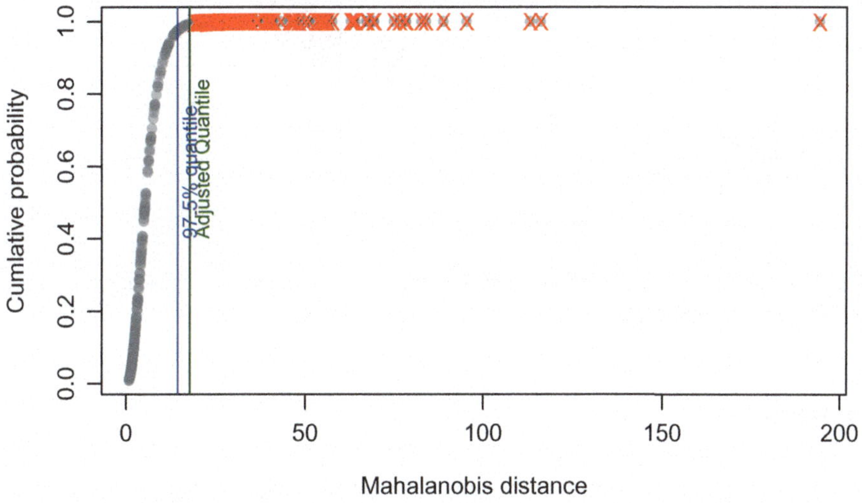

Fig. 7.10 Plot of the cumulative χ_m^2 function against Mahalanobis distance. The red crosses indicate possible outliers. The blue line is the 97.5% quantile, and the green line is the adjusted cutoff distance

We can now display (Fig. 7.11) the points selected as outliers using the adjusted cutoff value. As before, only three dimensions can be displayed, but the principal component space contains six dimensions.

```
# load the required package
require(scatterplot3d)

# plot the first three PCs along the identified outliers
sct3d <-scatterplot3d(pcspectraA$scores[,1],
                      pcspectraA$scores[,2],
                      pcspectraA$scores[,3],
                      xlab = "PC 1",
                      ylab = "PC 2",
                      zlab = "PC 3",
                      color = rgb(red = 0.5, green = 0.5, blue = 0.5, alpha = 0.5),
```

```
                    pch = 16,grid=TRUE, angle=50)

# add red cross to potential outliers
sct3d$points3d(pcspectraA$scores[,1][outliers],
                    pcspectraA$scores[,2][outliers],
                    pcspectraA$scores[,3][outliers],
                    pch = "X",
                    col = "red")
```

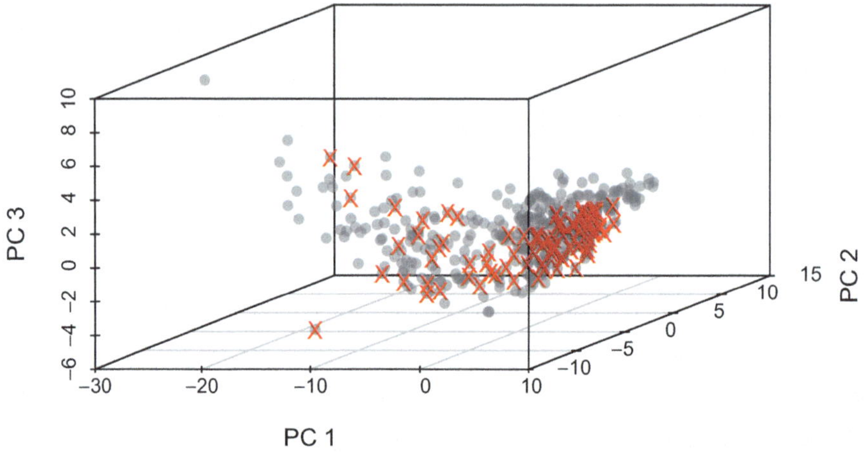

Fig. 7.11 Location of the potential outliers from the three first principal component score spaces of the absorbance spectra

References

Brereton RG (2003) Chemometrics: data analysis for the laboratory and chemical plant. Wiley, Chichester

Chang C-I (2000) An information-theoretic approach to spectral variability, similarity, and discrimination for hyperspectral image analysis. IEEE Trans Inf Theory 46:1927–1932

Chang C-I (2003) Hyperspectral imaging: techniques for spectral detection and classification. Springer Science & Business Media, Boston

Clark BJ, Frost T, Russell MA (1993) UV spectroscopy: techniques, instrumentation and data handling. Springer Science & Business Media, Berlin

Deza MM, Deza E (2009) Encyclopedia of distances. In: Encyclopedia of distances. Springer, Berlin/Heidelberg, pp 1–583

Du Y, Chang C-I, Ren H, Chang C-C, Jensen JO, D'Amico FM (2004) New hyperspectral discrimination measure for spectral characterization. Opt Eng 43:1777–1787

Farifteh J, Van Der Meer F, Carranza EJM (2007) Similarity measures for spectral discrimination of salt-affected soils. Int J Remote Sens 28:5273–5293

Filzmoser P, Garrett RG, Reimann C (2005) Multivariate outlier detection in exploration geochemistry. Comput Geosci 31:579–587

Garrett RG (1989) The chi-square plot: a tool for multivariate outlier recognition. J Geochem Explor 32:319–341

Kullback S, Leibler RA (1951) On information and sufficiency. Ann Math Stat 22:79–86

Mahalanobis PC (1936) On the generalized distance in statistics. In: Proceedings of the national institute of science, India. National Institute of Science of India
Mark H (1986) Normalized distances for qualitative near-infrared reflectance analysis. Anal Chem 58:379–384
Mark H, Workman Jr J (2010) Chemometrics in spectroscopy. Elsevier, Boston
Ramirez-Lopez L, Behrens T, Schmidt K, Viscarra-Rossel RA, Demattê JAM, Scholten T (2013) Distance and similarity-search metrics for use with soil vis–NIR spectra. Geoderma 199:43–53
Rousseeuw PJ, Driessen KV (1999) A fast algorithm for the minimum covariance determinant estimator. Technometrics 41:212–223
Schwarz J, Staenz K (2001) Adaptive threshold for spectral matching of hyperspectral data. Can J Remote Sens 27:216–224
Shenk JS, Westerhaus MO (1991) Population definition, sample selection, and calibration procedures for near infrared reflectance spectroscopy. Crop Sci 31:469–474
Varmuza K, Filzmoser P (2016) Introduction to multivariate statistical analysis in chemometrics. CRC Press, Boca Raton
Yuhas RH, Goetz AFH, Boardman JW (1992) Discrimination among semi-arid landscape endmembers using the spectral angle mapper (SAM) algorithm. In: Summaries 3rd annu. JPL airborne geoscience workshop, pp 147–149

Chapter 8
Selection of the Samples for Laboratory Analysis

Usually, spectra are obtained for all the soil samples available, but only a subset of these samples are sent to the laboratory for chemical and physical analysis. The reason is that spectra are fast and cheap to retrieve, while a single soil analysis (e.g. for soil clay) is relatively slow, and significantly more costly, to obtain. One must select a 'representative' subset of soil samples to be sent to the laboratory (Daszykowski et al. 2002). The spectra and the associated soil property values obtained by laboratory analysis are used to calibrate a regression model (see Chap. 9). The sampling design and sample size of the subsample to be sent for laboratory analysis will play a key role in the resulting calibrated model accuracy.

There are several available sampling designs to select a subset of representative soil samples using the spectra. In this chapter, we describe a few of them often used in spectroscopic research. Some references that describe such sampling designs in more detail include Daszykowski et al. (2002), De Gruijter et al. (2006) and Brus (2019). In theory, the sampling design can be based on the value of the property of interest (e.g. the soil clay content) or on the spectra. Usually, the value of the soil property is obtained from the spectra, and thus it is not realistic to generate a subsample based on the yet-to-be-determined response value. We therefore focus in this section on the selection of the calibration sample using the soil spectral data.

In addition to the sampling design, this chapter also considers the size of the subsample to be sent to the laboratory. Ideally, one would like to send the smallest number of soil samples, to save costs, without compromising model prediction accuracy. To select a subsample, it is necessary to assess the representativeness of this subsample. Several approaches have been developed, in particular by Ramirez-Lopez et al. (2014) where the representativity of the subsample is assessed by computing the difference between the probability density function of the subsample to that of the whole set of available spectra. We follow this approach in this chapter.

The sampling designs illustrated in this chapter are simple random sampling, Kennard-Stone, *k*-means clustering and conditioned Latin hypercube sampling. We further show the effect of outlier on selecting the sampling design. Finally, we use

A. M. J.-C. Wadoux et al., *Soil Spectral Inference with R*, Progress in Soil Science,
https://doi.org/10.1007/978-3-030-64896-1_8

the method proposed by Ramirez-Lopez et al. (2014) to select an optimal subsample size.

The set of packages used in this chapter are installed using the lines below. The book-associated `soilspec` package is also required; see Chap. 3 for information on its installation.

```
# specify all the packages used in the chapter and install them
myPackages <- c("matrixStats", "clhs", "viridis", "viridisLite",
                "prospectr", "RcppArmadillo", "scatterplot3d", "resemble")

# define which packages are not installed in the current computer
notInstalled <- myPackages[!(myPackages %in% installed.packages()[ , "Package"])]

# install the missing packages
if(length(notInstalled)) install.packages(notInstalled)
```

We use the raw spectra used in the previous chapters. These spectra are provided through the book-associated `soilspec` R package.

```
# load the package
require(soilspec)

# load the data
data("datsoilspc")

# convert the reflectance to absorbance spectra
spectraA <-  log(1/datsoilspc$spc)
```

To select a calibration sample, we transform the original spectra using a dimension reduction routine such as principal component analysis. Principal component analysis is presented in Sect. 6.2.

```
library(resemble)

# indicate the maximum amount of cumulative variance explained
# that needs to be retained in the PCs
maxexplvar <- 0.999

# perform PCA
pcspectraA <- pc_projection(Xr = spectraA,
                            pc_selection = list("cumvar", maxexplvar),
                            center = TRUE, scale = FALSE)


print(pcspectraA)
```

```
##
##  Method:  pca (svd)
##  Number of components retained:  9
##  Number of observations and number of original variables:  391 2151
##
##  Standard deviations, cumulative variance explained, individual variance explained:
##
##  Explained variance in X {Xr; Xu}:
##                             pc_1  pc_2   pc_3   pc_4    pc_5    pc_6  pc_7
## sd                        11.229 4.205 2.4079 1.2496 0.63800 0.50328 0.391
## cumulative_explained_var   0.829 0.945 0.9829 0.9932 0.99585 0.99751 0.999
```

```
## explained_var             0.829 0.116 0.0381 0.0103 0.00267 0.00166 0.001
##                               pc_8      pc_9
## sd                        0.201217 0.167825
## cumulative_explained_var 0.998782 0.998968
## explained_var             0.000266 0.000185
```

8.1 Sampling Design

From the sample of spectra of size n, the objective is to select a subsample, in our case 25, that will be sent to the laboratory for chemical and physical analysis.

```
SampleSize <- 25
```

8.1.1 *Simple Random Sampling*

Simple random sampling is the simplest form of sampling design where a sample of fixed size (here 25) is selected randomly from the population (all available spectra). No restriction is imposed other than the sample size. Simple random sampling is a probability sampling design in that the selection probability of each spectrum can be calculated beforehand. In simple random sampling, each unit has equal probability of being included in the sample (De Gruijter et al. 2006). When the sample is collected with replacement, a unit can be selected more than once in the sample. In this book, we use sampling without replacement, which means that each spectrum is unique in the subsample.

```
# set the seed for reproducibility
set.seed(19101991)

# select the row number using simple random sampling
randId <- sample(1:nrow(pcspectraA$scores),
                 size = SampleSize)

# load the required package
require(scatterplot3d)

# plot the first three PCs
sct3d <-scatterplot3d(pcspectraA$scores[,1],
                      pcspectraA$scores[,2],
                      pcspectraA$scores[,3],
                      xlab = "PC 1",
                      ylab = "PC 2",
                      zlab = "PC 3",
                      color= rgb(red = 0.5, green = 0.5, blue = 0.5, alpha = 0.5),
                      pch = 16, grid=TRUE, angle=50)

# plot the selected calibration sample
sct3d$points3d(pcspectraA$scores[,1][randId],
               pcspectraA$scores[,2][randId],
               pcspectraA$scores[,3][randId],
               pch = 1,
               col = "red",
               cex = 1.5)
```

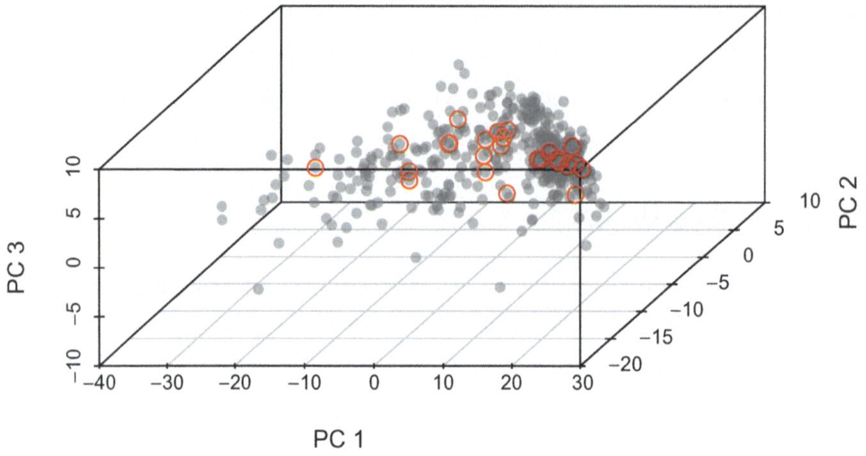

Fig. 8.1 Plot of the three first principal component scores (grey dots) and the selected sample of size 25 (red circles). The sample is selected with simple random sampling

It is important to note that by repeating the selection of the calibration subsample with simple random sampling, the algorithm will select a different subsample for each repetition. In R this can be avoided by setting the argument `set.seed()` before the selection of the sample.

The 3-D plot in Fig. 8.1 shows the spread of the selected subsample relative to the whole dataset. Ideally, the subsample covers the space of the whole dataset. The main disadvantage of simple random sampling is that the coverage of the whole dataset space might be poor, that is, the selected subsample is not representative. This is illustrated in Fig. 8.1, where the spread of the subsample is not uniform and some spaces are not covered.

8.1.2 Kennard-Stone

The Kennard-Stone algorithm is an iterative sample selection algorithm. In a first step, the algorithm takes the pair of points that are the furthest away from one another in the multivariate space of the principal component scores. The distance metrics often used are the Euclidean distance or, when working with correlated variables, the Mahalanobis distance (Mahalanobis 1936; see Chap. 7). The algorithm then selects sequentially the sampling points that maximize the distance metric between the points already selected. The process is repeated until the sample size is reached.

The algorithm has two main drawbacks. First, it will always select points that are far away from the average in the principal component space, and thus outlier detection as a prior step is almost always necessary (see also Sect. 8.1.5). Second, the algorithm is time-consuming because each allocation requires testing different combinations.

We use the `prospectr` package and the `kenStone` function to select a sample with Kennard-Stone.

```
# load the required package
require(prospectr)

# set the seed
set.seed(19101991)

# select a subsample of size 25 using the Mahalanobis distance as criterion
kssS <- kenStone(X = pcspectraA$scores,
                 k = SampleSize,
                 metric = "mahal",
                 .center = TRUE, .scale = FALSE)

# summary of the kssS object created with the kenStone function
str(kssS)
```

```
## List of 2
##  $ model: int [1:25] 376 372 261 120 168 67 264 206 361 247 ...
##  $ test : int [1:366] 1 2 3 4 5 6 7 8 9 10 ...
```

We can plot the space of the principal component scores and the associated selected subsample (Fig. 8.2).

```
# load the required package
require(scatterplot3d)

# plot the first three PCs
sct3d <-scatterplot3d(pcspectraA$scores[,1],
                      pcspectraA$scores[,2],
                      pcspectraA$scores[,3],
                      xlab = "PC 1",
                      ylab = "PC 2",
                      zlab = "PC 3",
                      color= rgb(red = 0.5, green = 0.5, blue = 0.5, alpha = 0.5),
                      pch = 16, grid=TRUE, angle=50)

# plot the selected calibration sample
sct3d$points3d(pcspectraA$scores[,1][kssS$model],
               pcspectraA$scores[,2][kssS$model],
               pcspectraA$scores[,3][kssS$model],
               pch = 1,
               col = "red",
               cex = 1.5)
```

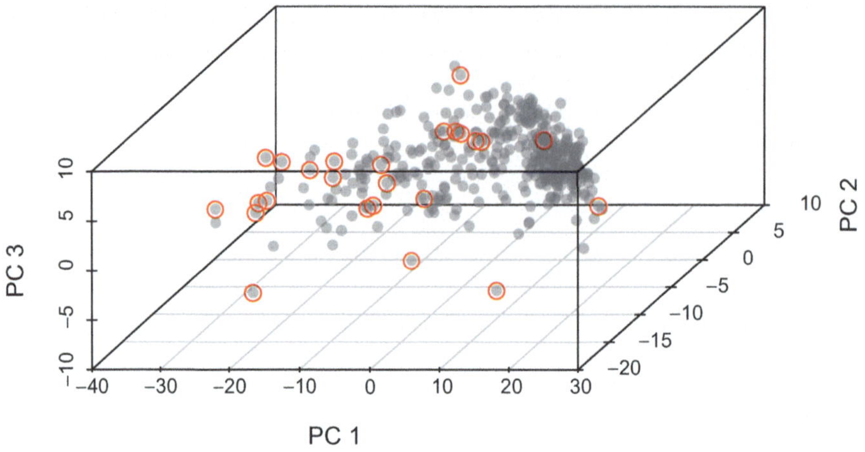

Fig. 8.2 Plot of the three first principal component score values (grey dots) and the selected subsample of size 25 (red circles). The sample is selected with Kennard-Stone. Note that a number of points appear on the peripheral area of this plot, rather than distributed throughout the population. This could be considered a drawback of using this algorithm for subsample selection

8.1.3 K-Means Clustering

K-means clustering is a very popular unsupervised classification algorithm. The underlying theory of this algorithm can be found in textbooks on multivariate data analysis (e.g. Hair et al. 2013) or machine learning. The algorithm starts by a random selection of the cluster centroids which are used as starting point to compute the clusters. The number of clusters is usually smaller than the sample size. For example, one might want to select more than one sample (at random) from each cluster, thereby reducing the overall cluster number. Whichever the case, the algorithm then performs an iterative process in which the cluster centroids are optimized to minimize an objective function. This objective function is the sum of square distances between the sample and the corresponding cluster centroids (MacQueen 1967). A number of distance metrics may be considered here depending on the data that is being clustered, including Euclidean or Mahalanobis distance which have been described previously.

Once the clusters have been defined, it is possible to take a point from each of them (if the sample size is equal to the number of clusters). This point is the closest to the centre of the cluster, randomly inside the cluster or the farthest away from the centre of the cluster.

We use the implementation provided by the `prospectr` package with the `naes` function. We set the argument `method = 0`, which means that we select the sample in the cluster centre. By setting the `method` argument to 1 or 2, we would select the sample the farthest away from the cluster centre or randomly inside the cluster, respectively (Fig. 8.3).

```
# load the required packages
require(prospectr)
require(viridis)

# set the seed
set.seed(19101991)

# we select the calibration sample with k-means clustering
# note that method = 1 takes the sample in the centre of the clusters
kmsS <- naes(X = pcspectraA$scores,
             k = SampleSize,
             method = 0,
             iter.max = 1000,
             .center = TRUE, .scale = FALSE)

# summary of the kmsS object created with the naes function
str(kmsS)
```

```
## List of 4
##  $ model   : int [1:25] 105 223 255 236 158 136 312 11 91 318 ...
##  $ test    : int [1:366] 1 2 3 4 5 6 7 8 9 10 ...
##  $ cluster: Named int [1:391] 8 11 7 16 2 16 25 25 12 18 ...
##   ..- attr(*, "names")= chr [1:391] "Xr_1" "Xr_2" "Xr_3" "Xr_4" ...
##  $ centers: num [1:25, 1:9] -5.7997 -15.0755 -0.0982 -27.2959 2.7034 ...
##   ..- attr(*, "dimnames")=List of 2
##   .. ..$ : chr [1:25] "1" "2" "3" "4" ...
##   .. ..$ : chr [1:9] "pc_1" "pc_2" "pc_3" "pc_4" ...
```

```
# load the required package
require(scatterplot3d)

# plot the first three PCs and use colours for the clusters
sct3d <-scatterplot3d(pcspectraA$scores[,1],
                      pcspectraA$scores[,2],
                      pcspectraA$scores[,3],
                      xlab = "PC 1",
                      ylab = "PC 2",
                      zlab = "PC 3",
                      color= viridis(25)[kmsS$cluster],
                      pch = 16, grid=TRUE, angle=50)

# plot the selected calibration sample
sct3d$points3d(pcspectraA$scores[,1][kmsS$model],
               pcspectraA$scores[,2][kmsS$model],
               pcspectraA$scores[,3][kmsS$model],
               pch = 1,
               col = "red",
               cex = 1.5)
```

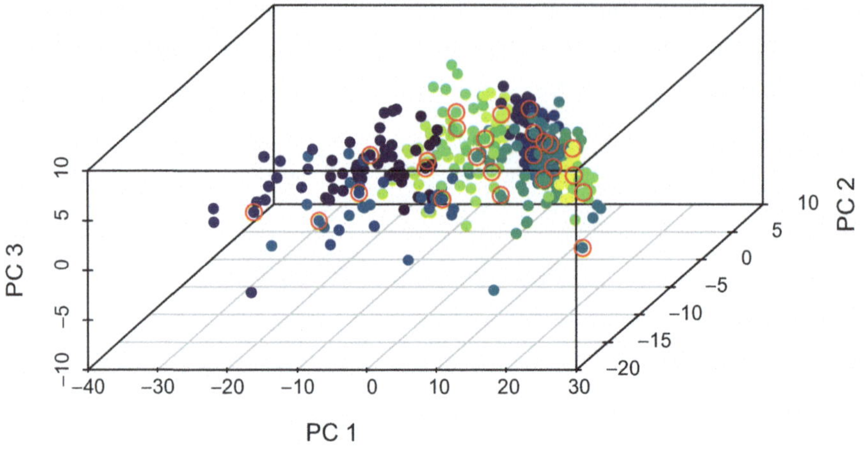

Fig. 8.3 Plot of the three first principal component score values (dots). The colour of each dot is according to the cluster each was assigned to using *k*-means clustering (here 25 classes). The dots with red circles are the selected samples which correspond to a sample from each cluster

8.1.4 Conditioned Latin Hypercube Sampling

Conditioned Latin hypercube (cLHS) is a sampling design which aims at covering equally the empirical distribution of a variable. For a single dimension, the cumulative distribution of the variable is divided into a number of strata whose number corresponds to the sample size. One single point is then selected in each stratum using simple random sampling. In a multivariate space, the sample is selected using an optimization algorithm to minimize an objective function which is a weighted sum of two components for quantitative variables. The first component called O_1 ensures that each variable contains one unit per stratum in the multi-dimensional feature space, while O_2 accounts for the correlation between the variable value in the sample and in the population. More information is found in Minasny and McBratney (2006).

Several implementations of cLHS are available in R. They are different in the way they compute the O_1 criterion so as to remove the dominance of O_1 over O_2 during the optimization. For example, the `spsann` package provides three implementations, each giving a different weight to the O_1 component. See also the description of the package. In this section, we provide an example based on the `clhs` package, which follows the implementation provided by the original publication of Minasny and McBratney (2006). Note however that the two components O_1 and O_2 will have different scale and the optimization algorithm will mostly reduce the value of the O_1 component (Fig. 8.4). More discussion of the differences between cLHS algorithms is provided by Brus (2019) and Wadoux et al. (2019).

```
# load the required package
require(clhs)

## set the seed
set.seed(19101991)

## since the clhs function accepts only 'data.frame' objects as input variables, we can
## transform our matrix of scores to 'data.frame' using the as.data.frame function
clhsS <- clhs(x = as.data.frame(pcspectraA$scores),
              size = SampleSize,
              iter = 1000,
              simple = FALSE)

## let ensure visually that the objective function was correctly minimized
plot(clhsS)
```

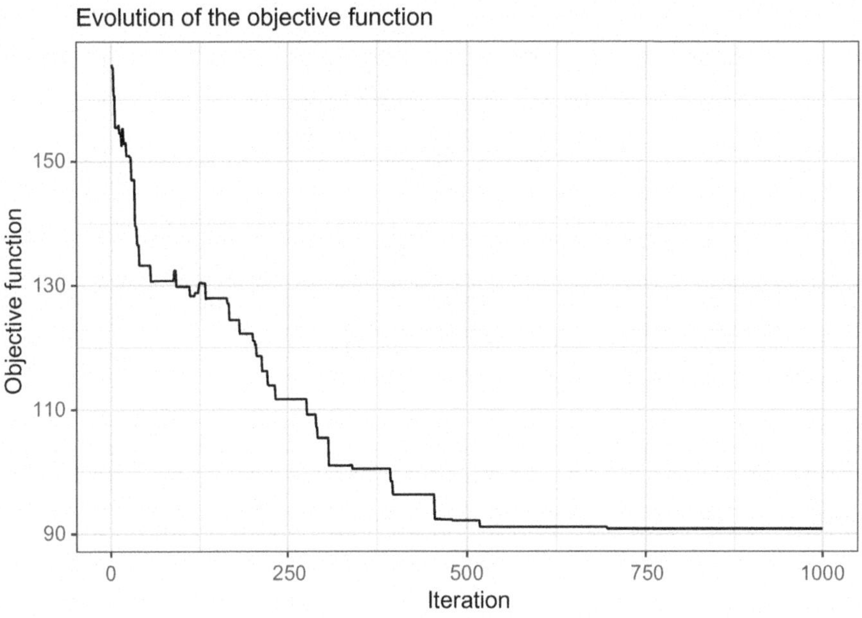

Fig. 8.4 Evolution of the cLHS objective function minimization

```
# summary of the clhs object created with the clhs function
str(clhsS)
```

We can now plot the selected sample (Fig. 8.5).

```
# load the required package
require(scatterplot3d)

# plot the first three PCs
sct3d <-scatterplot3d(pcspectraA$scores[,1],
                      pcspectraA$scores[,2],
                      pcspectraA$scores[,3],
```

```
                      xlab = "PC 1",
                      ylab = "PC 2",
                      zlab = "PC 3",
                      color= rgb(red = 0.5, green = 0.5, blue = 0.5, alpha = 0.5),
                      pch = 16, grid=TRUE, angle=50)

# plot the selected calibration sample
sct3d$points3d(pcspectraA$scores[,1][clhsS$index_samples],
               pcspectraA$scores[,2][clhsS$index_samples],
               pcspectraA$scores[,3][clhsS$index_samples],
               pch = 1,
               col = "red",
               cex = 1.5)
```

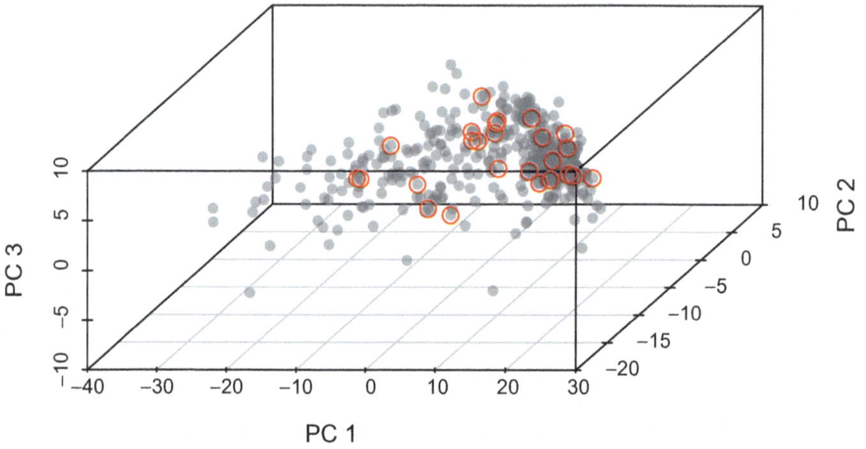

Fig. 8.5 Plot of the three first principal component score values (grey dots) and the selected sample of size 25 (red circles). The sample is selected with conditioned Latin hypercube sampling

8.1.5 Presence of Outliers in the Data

In the examples presented in the previous section, it is not obvious how well the different sampling designs perform in the presence of outliers in the data. Other than some purely qualitative assessments, the plot in three dimensions makes it also difficult to see where the sampling points are selected. We present here a two-dimensional case based on synthetic data to highlight the differences between designs.

Let us create some synthetic data (Fig. 8.6): a space with two variables, called `x1` and `x2`. Each variable ranges from 1 to 30. The first grid does not contain outliers. In the second grid, an outlier is positioned at the coordinates `c(35, 35)` of the two-dimensional space.

```
# create two sequences of number between 1 and 30
x1 <- seq(from = 1, to = 30, by = 1)
x2 <- seq(from = 1, to = 30, by = 1)

# make a grid from the generated data
griddata <- expand.grid(x1 = x1, x2 = x2)

# add random noise
griddata[,1] <- griddata[,1] + rnorm(length(griddata[,1] ), 0, 0.05)
griddata[,2] <- griddata[,2] + rnorm(length(griddata[,2] ), 0, 0.05)

# add an outlier
griddata <- rbind(c(35,35), griddata)

# plot the grid without and with the outlier
par(mfrow=c(1,2))

# plot the first grid without the outlier
plot(griddata$x1[-1], griddata$x2[-1],
     pch = 16,
     xlab = "x1",
     ylab = "x2",
     col = rgb(red = 0.5, green = 0.5, blue = 0.5, alpha = 0.5))


# plot the second grid with the outlier
plot(griddata$x1, griddata$x2,
     pch = 16,
     xlab = "x1",
     ylab = "x2",
     col = rgb(red = 0.5, green = 0.5, blue = 0.5, alpha = 0.5))
```

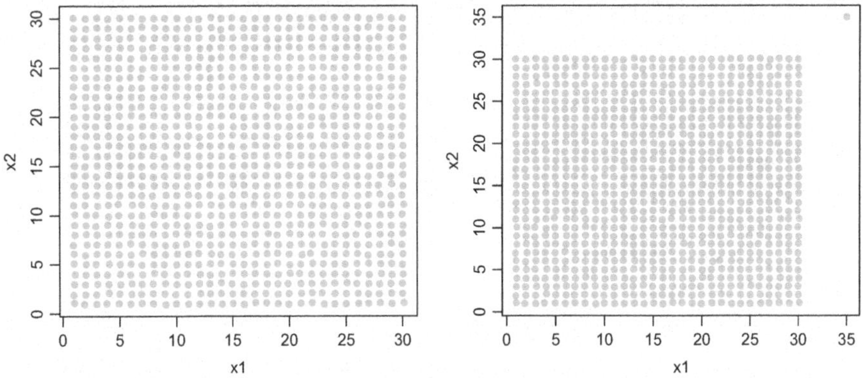

Fig. 8.6 Plot of the synthetic data generated without (left) or with an outlier (right)

```
# create a copy of the grid without the outlier
griddataNo <- griddata[-1,]
```

Now we can select a sample with each of the sampling designs that were examined previously. We can plot each of them for both cases, with or without the outlier (Fig. 8.7).

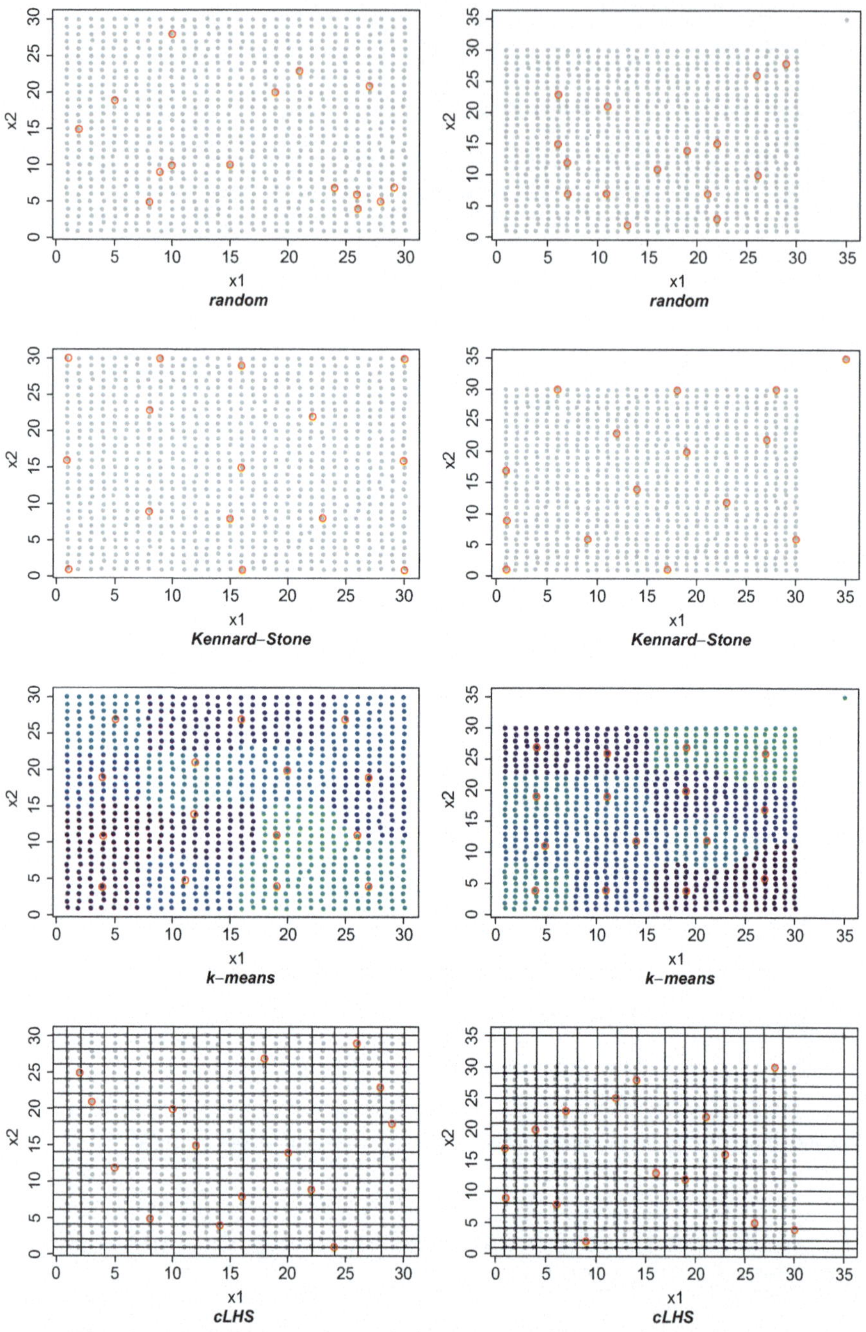

Fig. 8.7 Plot of the synthetic data generated without (left column) or with an outlier (right column) and with the selected samples in red. The sample size is 15. The samples are selected using simple random sampling (first row), Kennard-Stone (second row), *k*-means clustering (third row) and conditioned Latin hypercube sampling (fourth row). In the third row, the colours indicate the clusters. In the fourth row, the horizontal and vertical black lines indicate the breaks of the multivariate strata

Figure 8.7 shows that Kennard-Stone sampling is particularly sensitive to the outliers in the data as it selects the points that are the furtherest away. Simple random sampling did not select the outlier, but due to its randomness, it may well happen if we select another sample with a different seed. Both k-means sampling and cLHS did not select the outlier. For k-means, the cluster centre of the case with outlier is similar to that without outlier. In cLHS, the sampling in the multivariate stratum covering the outlier is covered by sampling in another stratum.

8.2 Sample Size

8.2.1 Assessing the Representativeness of the Sample

To select an optimal sample size, Ramirez-Lopez et al. (2014) (and others, e.g. Stumpf et al. 2016) suggested comparing the selected sample and the whole population on the basis of the probability density function computed on the scores of the principal components. A sample of a given size is representative of the whole population if the density of the sample is similar to that of the whole population. To reduce dimensionality, the density is computed on the principal component scores.

Let us take an example with the `NIRsoil` data from `prospectr`.

```
# load the required packages
require(prospectr)
require(resemble)

# load the data
data(NIRsoil)

# take the spectra (stored in a matrix called 'spc').
spectraA <- NIRsoil[,c("spc")]

wavs <- as.numeric(colnames(spectraA))

# choose the maximum cumulative variance explained
maxexplvar <- 0.999

# make the PCA
pcspectra <- pc_projection(Xr = spectraA,
                           pc_selection = list("cumvar", maxexplvar),
                           center = TRUE, scale = FALSE)

# print the results of the PCA
print(pcspectra)
```

```
##
## Method: pca (svd)
## Number of components retained: 3
## Number of observations and number of original variables: 825 700
##
## Standard deviations, cumulative variance explained, individual variance explained:
##
## Explained variance in X {Xr; Xu}:
##                         pc_1   pc_2    pc_3
## sd                      2.260 0.3041 0.10320
```

```
## cumulative_explained_var 0.978 0.9960 0.99806
## explained_var            0.978 0.0177 0.00204
```

We take three samples from the whole population using simple random sampling.

```
# set the seed
set.seed(19101991)

# select a sample of size 100 using simple random sample
sampleA <- sample(1:nrow(pcspectra$scores), size = 100)
sampleB <- sample(1:nrow(pcspectra$scores), size = 100)
sampleC <- sample(1:nrow(pcspectra$scores), size = 100)
```

Now the density can be derived using the `density` function from the `stats` package.

```
# density of the scores for the whole population
densAll <- density(pcspectra$scores)$y

# density of the scores for the samples
densSampleA <- density(pcspectra$scores[sampleA,])$y
densSampleB <- density(pcspectra$scores[sampleB,])$y
densSampleC <- density(pcspectra$scores[sampleC,])$y

# plot the density of the population
par(mfrow=c(1,3))
plot(densAll,
     type = "l",
     xlab = "Index",
     ylab = "Density",
     main = "Sample A",
     col = rgb(red = 0.1, green = 0.1, blue = 0.1, alpha = 1))

# Add the line of the sample A density
lines(densSampleA,
      col = rgb(red = 1, green = 0, blue = 0, alpha = 1))

# plot the density of the population
plot(densAll,
     type = "l",
     xlab = "Index",
     ylab = "Density",
     main = "Sample B",
     col = rgb(red = 0.1, green = 0.1, blue = 0.1, alpha = 1))

# Add the line of the sample B density
lines(densSampleB,
      col = rgb(red = 0, green = 1, blue = 0, alpha = 1))

# plot the density of the population
plot(densAll,
     type = "l",
     xlab = "Index",
     ylab = "Density",
     main = "Sample C",
     col = rgb(red = 0.1, green = 0.1, blue = 0.1, alpha = 1))

# Add the line of the sample C density
```

```
lines(densSampleC,
      col = rgb(red = 0, green = 0, blue = 1, alpha = 1))
```

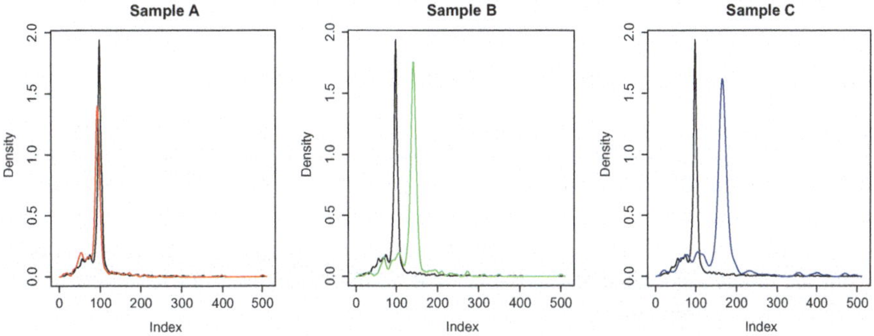

Fig. 8.8 Density of the principal component scores for three selected samples (in colour) compared to the density of the whole dataset principal component scores (in black). The samples are selected by simple random sampling

Figure 8.8 shows that the density of sample A is very close to that of the whole population, while densities of samples B and C are quite different.

Ramirez-Lopez et al. (2014) quantified the similarity between probability density functions (pdfs) by computing the mean square Euclidean distance (msd) between the pdfs. Probably a more standard approach for comparing empirical distribution functions is the Kullback & Leibler (also called relative entropy) divergence (Kullback and Leibler 1951) which is specifically designed for comparing distributions. For the Ramirez-Lopez et al. (2014) approach, the msd is computed as the mean squared Euclidean distance (see Sect. 7.1.1) between estimates of the pdfs of the principal component scores and that of the subsample.

This is computed as follows:

```
# mean square distance between the population and sample pdfs
# sample A
mean((densAll - densSampleA)^2, na.rm = T)
```

```
## [1] 0.01382066
```

```
# sample B
mean((densAll - densSampleB)^2, na.rm = T)
```

```
## [1] 0.09982772
```

```
# sample C
mean((densAll - densSampleC)^2, na.rm = T)
```

```
## [1] 0.1170962
```

The values confirm the visual assessment, i.e. the pdf of sample A is closer than the pdf of the population, while the pdfs of samples B and C are the second and third closest from the pdf of the population. This means that sample A is more representative of the whole population than sample B or C. For a calibration exercise, we would therefore favour (sub)sample A over B and C.

8.2.2 Optimal Sample Size Based on the Spectra

Now that we know how to assess the representativity of a sample, we can perform this analysis for different sampling designs and sample sizes. In R it would require to make a nested `for` loop. Instead, we build a function that can be loaded from the book-associated `soilspec` package. This function is called `css()` and comes from the codes associated to the paper of Ramirez-Lopez et al. (2019). The codes are freely accessible online.

```
# load the required package
require(soilspec)
```

This function enables to compare three different designs (viz. Kennard-Stone, *k*-means clustering and cLHS) for different user-defined sample sizes.

```
# display the arguments of the css function
args(css)
```

```
## function (S, k, method = "kms", repetitions = 10, n = 512, from,
##     to, bw, ...)
## NULL
```

- `S`: A matrix of the scores of the principal components.
- `k`: A vector containing the sample set sizes to be evaluated.
- `method`: The sampling algorithm. Options are (1) `"kss"` (Kennard-Stone sampling); (2) `"kms"` (*k*-means sampling), the default; and (3) `"clhs"` (conditioned Latin hypercube sampling).
- `repetitions`: The number of times that the sampling must be carried out for each sample size to be evaluated. The result of the final msd is the average of the ones obtained at each iteration. Note that since the `"kss"` method is deterministic and always returns the same results, there is no need for repetitions.
- `n`: The number of equally spaced points at which the probability densities are to be estimated (see `density` function of the package `stats`).
- `from`, `to`: A vector of the left- and right-most points of the grid at which the densities are to be estimated. Default is the minimums and maximums of the variables in `S`.
- `bw`: A vector containing the smoothing bandwidth to be used for the probability densities (see `density` function of the package `stats`).
- `...`: arguments to be passed to the calibration sampling algorithms, i.e. additional arguments to be used for the `clhs`, `kenStone` or `naes` functions which run inside this function.

We can use this function to compare the probability density functions of the population to that of the sample selected by the designs, for different sample sizes.

We start by standardizing (zero mean and unit variance) the principal component scores from the NIRsoil data that we used at the previous section.

```
# scale the PC scores
scaledPcs <- scale(pcspectra$scores, center = TRUE, scale = TRUE)

# load the required package
require(matrixStats)

# show the standard deviation of the columns
colSds(scaledPcs)
```

```
## [1] 1 1 1
```

```
# show the mean of the columns
colMeans(scaledPcs)
```

```
##          pc_1          pc_2          pc_3
##  5.925395e-18  6.335420e-18 -8.061207e-18
```

We then define a vector indicating the different sample sizes to be evaluated.

```
# sequence of sample size from 30 up to 360 in step of 30
sss <- seq(from = 30, to = 360, by = 30)
```

We can now use the css function to evaluate different sample size (Fig. 8.9 and 8.10).

```
# run the css function for Kennard-Stone sampling design
ksSs <- css(S = scaledPcs,
            k = sss,
            method = "kss",
            repetitions = 1)

# show the results
print(ksSs)
```

```
##    css         msd
## 1   30 0.029741853
## 2   60 0.021756504
## 3   90 0.017475285
## 4  120 0.014930065
## 5  150 0.014494793
## 6  180 0.013138603
## 7  210 0.011974236
## 8  240 0.010896802
## 9  270 0.010035836
## 10 300 0.008964971
## 11 330 0.008193682
## 12 360 0.007538720
```

```
# plot the results
plot(x = ksSs$css,
     y = ksSs$msd,
     xlab = "Sample size",
     ylab = "msd",
     type = "b",
     col = rgb(red = 0, green = 0.4, blue = 0.8, alpha = 0.5))
```

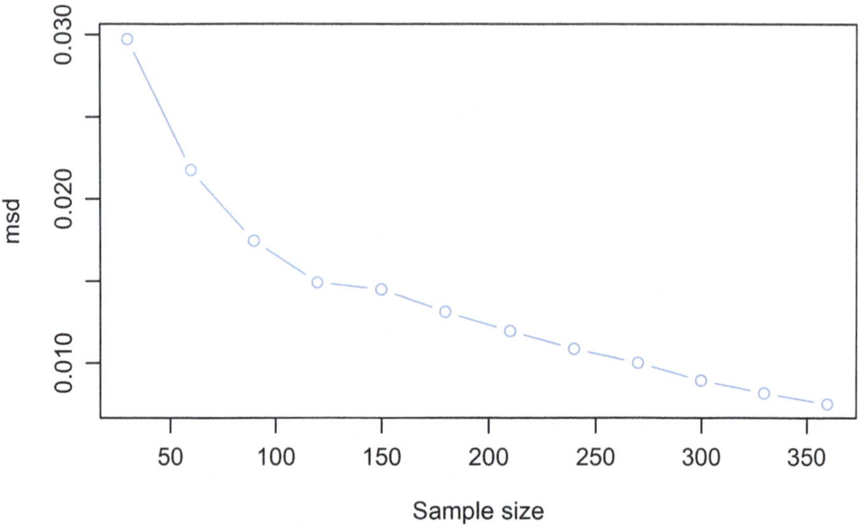

Fig. 8.9 Values of the mean squared Euclidean distance between the probability density function of the sample and that of the population. The samples are selected by Kennard-Stone sampling

```
# run the css function for a sample selected by k-means clustering
kmSs <- css(S = scaledPcs,
            k = sss,
            method = "kms",
            repetitions = 3)

# show the results
print(kmSs)
```

```
##     css          msd       msd_sd
## 1    30 0.0062405805 0.0035807958
## 2    60 0.0024765784 0.0014263296
## 3    90 0.0016699215 0.0009625250
## 4   120 0.0007047638 0.0004066092
## 5   150 0.0006171237 0.0003560769
## 6   180 0.0008057661 0.0004648350
## 7   210 0.0005004857 0.0002888111
## 8   240 0.0005580265 0.0003219971
```

```
## 9   270 0.0003103580 0.0001791297
## 10  300 0.0003981636 0.0002297884
## 11  330 0.0002100591 0.0001212522
## 12  360 0.0002345960 0.0001354123
```

```
# plot the results
plot(x = kmSs$css,
     y = kmSs$msd,
     xlab = "Sample size",
     ylab = "msd",
     type = "b",
     col = rgb(red = 0, green = 0.4, blue = 0.8, alpha = 0.5))
```

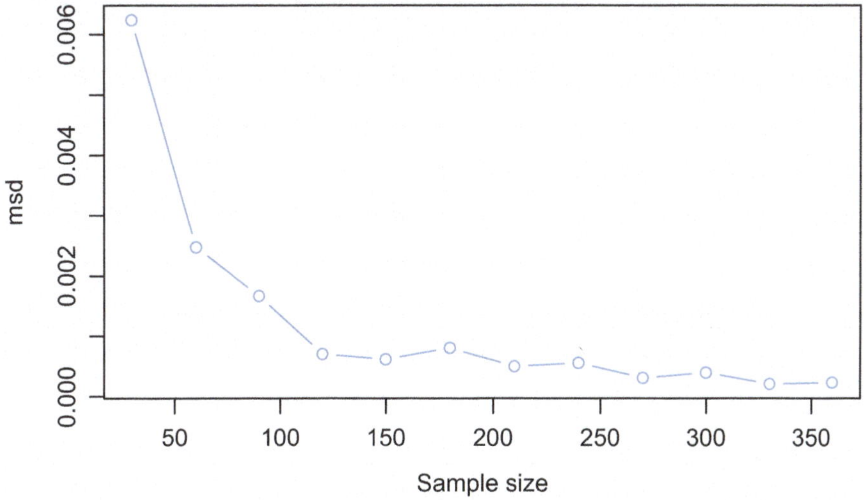

Fig. 8.10 Values of the mean squared Euclidean distance between the probability density function of the sample and that of the population. The samples are selected by k-means clustering

We use it also for cLHS (Fig. 8.11), but this time we pass the `iter` argument to the `clhs` function (which is executed internally) to set the number of iterations to 1,000 (the default of the `clhs` function is 10,000, but for the example, we set it to 1,000). To avoid convergence issues, we recommend to have a large number of iterations in the annealing schedule.

```
# run the css function for a sample selected by clhs
lhSs <- css(S = scaledPcs,
            k = sss,
            method = "clhs",
            repetitions = 3,
            iter = 1000)

# show the results
print(lhSs)
```

```
##      css          msd       msd_sd
## 1     30 0.0009644352 5.562810e-04
## 2     60 0.0006845639 3.949629e-04
## 3     90 0.0006694690 3.862597e-04
## 4    120 0.0007448093 4.296961e-04
## 5    150 0.0004972384 2.869381e-04
## 6    180 0.0002391695 1.380515e-04
## 7    210 0.0002829130 1.632937e-04
## 8    240 0.0001901029 1.097351e-04
## 9    270 0.0002247748 1.297447e-04
## 10 300 0.0001366639 7.889217e-05
## 11 330 0.0001687036 9.738467e-05
## 12 360 0.0001185249 6.842230e-05
```

```
# plot the results
plot(x = lhSs$css,
     y = lhSs$msd,
     xlab = "Sample size",
     ylab = "msd",
     type = "b",
     col = rgb(red = 0, green = 0.4, blue = 0.8, alpha = 0.5))
```

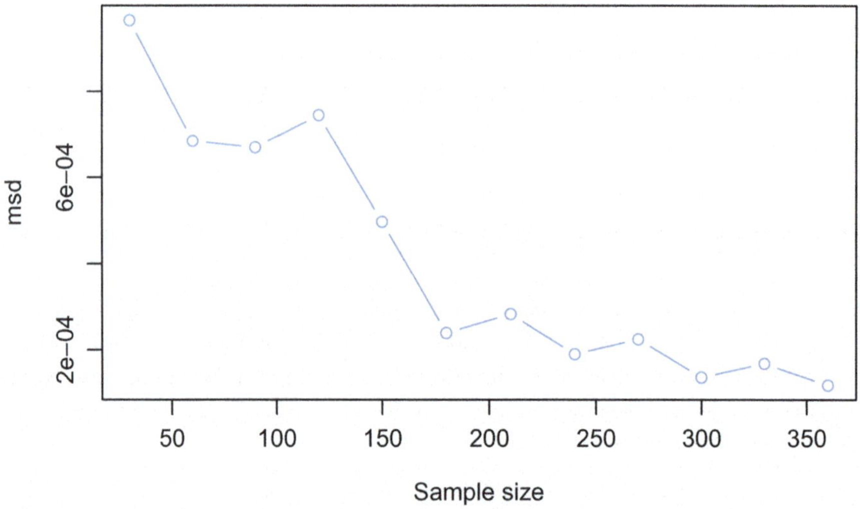

Fig. 8.11 Values of the mean squared Euclidean distance between the probability density function of the sample and that of the population. The samples are selected by conditioned Latin hypercube sampling

We now can plot the three results in one single plot (Fig. 8.12).

```
# plot all the three results in one graph
plot(x = ksSs$css,
     y = ksSs$msd,
     ylim = c(0, 0.03),
     xlab = "Sample size",
     ylab = "msd",
     type = "b",
     col = rgb(red = 1, green = 0, blue = 0, alpha = 0.9))

# add the points corresponding to the k-means sampling
points(x = kmSs$css,
       y = kmSs$msd,
       type = "b",
       col = rgb(red = 0, green = 1, blue = 0, alpha = 0.9))

# add the points corresponding to the conditioned Latin hypercube
points(x = lhSs$css,
       y = lhSs$msd,
       type = "b",
       col = rgb(red = 0, green = 0, blue = 1, alpha = 0.9))

# add a legend
legend(150, 0.030,
       legend = c("Kennard-Stone", "k-means", "conditioned Latin Hypercube"),
       col = c("red","green" ,"blue"),
       lty = 4,
       cex = 1)
```

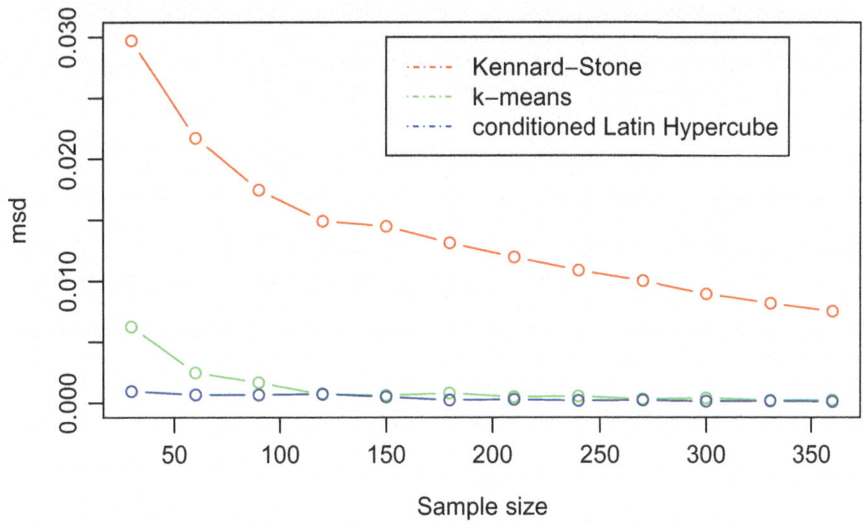

Fig. 8.12 Values of the mean squared Euclidean distance between the probability density function of the sample and that of the population. The samples are selected by Kennard-Stone sampling, *k*-means clustering or conditioned Latin hypercube sampling

Figure 8.12 shows that for larger sample sizes, the difference in terms of Euclidean distance between the probability density function of the sample and that of the population decreases. In all cases, sampling with Kennard-Stone yields the largest differences, while they are relatively small for cLHS and sampling with k-means clustering. Sampling with k-means clustering offers a good compromise between speed of the algorithm and small difference between the sample and population pdfs. Note that these results are case-specific and that for each spectral library one should carry out this experiment to assess the optimal sample size and sampling design to select them.

References

Brus DJ (2019) Sampling for digital soil mapping: a tutorial supported by R scripts. Geoderma 338:464–480

Daszykowski M, Walczak B, Massart DL (2002) Representative subset selection. Anal Chim Acta 468:91–103

De Gruijter JJ, Brus DJ, Bierkens MFP, Knotters M (2006) Sampling for natural resource monitoring. Springer Science & Business Media, Dordrecht

Hair JF, Black WC, Babin BJ, Anderson RE (2013) Multivariate data analysis. Pearson Education Limited, Upper Saddle River

Kullback S, Leibler RA (1951) On information and sufficiency. Ann Math Stat 22:79–86

MacQueen J (1967) Some methods for classification and analysis of multivariate observations. In: Proceedings of the fifth Berkeley symposium on mathematical statistics and probability, volume 1: statistics. University of California Press, Berkeley, pp 281–297

Mahalanobis PC (1936) On the generalized distance in statistics. In: Proceedings of the national institute of science, India. National Institute of Science of India

Minasny B, McBratney AB (2006) A conditioned Latin hypercube method for sampling in the presence of ancillary information. Comput Geosci 32:1378–1388

Ramirez-Lopez L, Schmidt K, Behrens T, Van Wesemael B, Demattê JA, Scholten T (2014) Sampling optimal calibration sets in soil infrared spectroscopy. Geoderma 226:140–150

Ramirez-Lopez L, Wadoux AMJ-C, Franceschini MHD, Terra FS, Marques KPP, Sayão VM, Demattê JAM (2019) Robust soil mapping at the farm scale with vis–NIR spectroscopy. Eur J Soil Sci 70:378–393

Stumpf F, Schmidt K, Behrens T, Schönbrodt-Stitt S, Buzzo G, Dumperth C, Wadoux A, Xiang W, Scholten T (2016) Incorporating limited field operability and legacy soil samples in a hypercube sampling design for digital soil mapping. J Plant Nutr Soil Sci 179:499–509

Wadoux AMJ-C, Brus DJ, Heuvelink GBM (2019) Sampling design optimization for soil mapping with random forest. Geoderma 355:113913

Chapter 9
Estimating Soil Properties and Classes from Spectra

The most common way of estimating soil properties from pre-processed spectra is by calibrating a statistical model. If the response of the spectra at a particular wavelength follows the Beer-Lambert law, the degree of reflectance at a particular wavelength is proportional to the concentration of a soil property. In this case, a linear model can be fitted between this wavelength and the measured values of a soil property. In most cases, however, the response of the spectrum follows a complex form, i.e. the concentration of a soil property is related to several interacting wavelengths and overlapping regions of the spectrum. In recent years, chemometric methods based on multivariate statistical models and machine learning algorithms have considered the entire spectrum as a predictor. When there are many hundreds of predictor variables (wavelengths), the methods can be described as multivariate. Multivariate models can be calibrated using the whole spectrum, with the target variables being measured values of soil properties.

Predictions made by a calibrated model need to be validated. Three common validation methods exist: data splitting, cross-validation and additional probability sampling. In digital soil spectroscopy, one most often has only a single dataset for both calibration and validation. Collecting an additional probability sample to independently validate soil spectral models is generally not feasible. Both data splitting and cross-validation are sub-optimal compared to collecting an additional probability sample because the information contained in the dataset cannot be fully exploited during calibration. In most cases, unfortunately, this is the only option.

In data splitting, the dataset is split into two subsets, generally containing 75% and 25% of the data, which are used for calibration and validation, respectively. In cross-validation, the dataset is split into K subsamples, where the $K-1$ subsamples are used for calibration, and validation statistics are computed from the subsample left aside. Each soil sample is used once for validation. Cross-validation should be preferred over simple data splitting, especially when the dataset is small (Brus et al. 2011).

A. M. J.-C. Wadoux et al., *Soil Spectral Inference with R*, Progress in Soil Science,
https://doi.org/10.1007/978-3-030-64896-1_9

In this chapter, we use data-splitting for the sake of demonstration. More information on validation methods can be found in the statistical (e.g. Friedman et al. 2001, Chapter 7) or pedometrics (e.g. Brus et al. 2011) literature.

The set of packages used in this chapter are installed using the lines below. The book-associated `soilspec` package is also required; see Chap. 3 for information on its installation.

```
# specify all the packages used in the chapter and install them if they are not already
myPackages <- c("caret", "ggplot2", "soiltexture", "resemble",
                "randomForest", "Cubist", "lattice", "pls",
                "prospectr", "RcppArmadillo")

# define which packages are not installed in the current computer
notInstalled <- myPackages[!(myPackages %in% installed.packages()[ , "Package"])]

# install the missing packages
if(length(notInstalled)>0) install.packages(notInstalled)
```

9.1 Goodness of Fit Measures

The process of creating a model begins first with a calibration. After a model is calibrated, we can use it to make a prediction on new samples. An important step in the analysis is to evaluate the calibrated model by predicting the value of the soil property and to comparing it with its associated measured value.

In the following sections, we will review the most common indicators of the quality of a prediction made by a calibrated model. These indicators are routinely employed in soil spectroscopy and in the general statistical modelling literature.

Root mean square error (RMSE) The root mean square error (RMSE) is the standard deviation of the residuals between observed and predicted values of a variable. The RMSE evaluates the dispersion of the residuals. In other words, the RMSE tells how concentrated the data are around the line of the best fit between observed and predicted values. RMSE values are non-negatives; a value of 0 indicates a perfect fit between observed and predicted values. The RMSE value depends on the scale of the data and is therefore not suitable for comparison between datasets. In general, the lower the RMSE, the better the fit between observed and predicted values. The RMSE is computed as follows:

$$\text{RMSE} = \sqrt{\frac{1}{n}\sum_{i=1}^{n}(\text{obs}_i - \text{pred}_i)^2}, \tag{9.1}$$

where n is the validation sample size and obs and pred are vectors of observed and predicted values of the soil properties, respectively. The RMSE can simply be derived in R by:

```
RMSE <- function(obs, pred){
  sqrt(mean((pred - obs)^2, na.rm = TRUE))
}
```

Mean error (ME) or bias The mean error (ME) is often computed along with the RMSE to assess the bias of the predictions. The ME is simply the average of all the errors between predictions and observations. Ideally, the value of the ME is zero which indicates no bias in the prediction. Note that a ME of zero does not indicate that there is no error (the positive and negative errors cancel out), but that there is no systematic bias in the predictions made by the model. The ME is computed as follows:

$$\mathrm{ME} = \frac{1}{n}\sum_{i=1}^{n}(\mathrm{obs}_i - \mathrm{pred}_i), \tag{9.2}$$

which in R gives:

```
ME <- function(obs, pred){
  mean(pred - obs, na.rm = TRUE)
}
```

Squared correlation coefficient (r^2) Pearson's squared correlation coefficient (r^2) is commonly used to assess the dispersion around the regression line. In other words, the r^2 represents the strength of the linear association between observed and predicted values with respect to the fitted regression line. The r^2 is such that the values are between 0 and 1. It is computed as follows:

$$r^2 = \left(\frac{\sum_{i=1}^{n}(\mathrm{obs}_i - \overline{\mathrm{obs}})(\mathrm{pred}_i - \overline{\mathrm{pred}})}{\sqrt{\sum_{i=1}^{n}(\mathrm{obs}_i - \overline{\mathrm{obs}})^2}\sqrt{\sum_{i=1}^{n}(\mathrm{pred}_i - \overline{\mathrm{pred}})^2}}\right)^2. \tag{9.3}$$

In R, this can be efficiently derived using the `cor` function from the `stats` package:

```
r2 <- function(obs, pred){
  cor(pred, obs, method = "spearman", use = "pairwise.complete.obs")^2
}
```

While the r^2 is widely used, there is a general confusion in the literature about what a r^2 is and how to compute it. When the r^2 is computed as the squared Pearson's r correlation coefficient, it measures the closeness of fit to the fitted linear regression line between observed and predicted, but does not indicate the closeness against a 1:1 line (observed versus predicted) which is of interest when validating. The r^2 is not sensitive to the departure of fitted regression line to the 45 degree line of agreement. In many cases, it is therefore not recommended to compute the r^2 as the squared correlation coefficient, in particular when predictions are biased.

Coefficient of determination (R2) The coefficient of determination (R2) is the amount of variance explained by the model. The R2 quantifies the improvement made by the model over simply using the mean of the observations as prediction (Janssen and Heuberger 1995). In the literature, the R2 is sometimes referred to as the amount of variance explained, a modelling efficiency coefficient (Wadoux et al. 2018), a skill score (Nussbaum et al. 2017) or a Nash-Sutcliffe model efficiency coefficient (Nash and Sutcliffe 1970). As for the r^2, its optimal value is 1, but it can be negative if the root mean square error exceeds the standard deviation of the data. It is computed as follows:

$$\mathrm{R2} = 1 - \frac{\sum_{i=1}^{n}(\mathrm{obs}_i - \mathrm{pred}_i)^2}{\sum_{i=1}^{n}(\mathrm{obs}_i - \overline{\mathrm{obs}})^2} \tag{9.4}$$

which is equal to $1 - SSE/SST$ where SSE is the sum of the squared error and SST of the total sum of squares. The R2 is derived in R by:

```
R2 <- function(obs, pred){
  # sum of the squared error
  SSE <- sum((pred - obs) ^ 2, na.rm = T)
  # total sum of squares
  SST <- sum((obs - mean(obs, na.rm = T)) ^ 2, na.rm = T)
  R2 <- 1 - SSE/SST
  return(R2)
}
```

Lin's concordance coefficient (ρ_c) The concordance correlation coefficient (ρ_c) was introduced by Lawrence and Lin (1989) to assess the agreement between observed and predicted values with respect to the 1:1 line. If the predictions are in perfect agreement with the observations, all the points fall on the 1:1 line. The ρ_c is given by:

$$\rho_c = \frac{2r\sigma_{\mathrm{pred}}\sigma_{\mathrm{obs}}}{\sigma_{\mathrm{obs}}^2 + \sigma_{\mathrm{pred}}^2 + (\mu_{\mathrm{obs}} - \mu_{\mathrm{pred}})^2} = rC_b, \tag{9.5}$$

where r is Pearson's correlation coefficient, σ is the standard deviation ($r\sigma_{\mathrm{pred}}\sigma_{\mathrm{obs}}$ is the covariance between observed and predicted values) and μ is the mean. Lawrence and Lin (1989) have shown that ρ_c reduces to rC_b where C_b is the bias correction factor defined as:

$$C_b = \left(\frac{v + 1/v + u^2}{2}\right)^{-1}, \tag{9.6}$$

with $v = \sigma_{\mathrm{pred}}/\sigma_{\mathrm{obs}}$ being the scale shift and $u = (\mu_{\mathrm{pred}} - \mu_{\mathrm{obs}})/\sqrt{\sigma_{\mathrm{pred}}\sigma_{\mathrm{obs}}}$ being the location shift relative to the scale. In other terms, ρ_c assesses the correlation between observed and predicted values, with a bias correction. The ρ_c can be implemented in R by:

```
rhoC <- function(obs, pred) {
  n <- length(pred)
  sdPred <- sd(pred, na.rm = T)
  sdObs <- sd(obs, na.rm = T)
  r <- stats::cor(pred, obs, method = "pearson", use = "pairwise.complete.obs")
  # scale shift
  v <- sdPred / sdObs
  sPred2 <- var(pred, na.rm = T) * (n - 1) / n
  sObs2 <- var(obs, na.rm = T) * (n - 1) / n
  # location shift relative to scale
  u <- (mean(pred, na.rm = T) - mean(obs, na.rm = T)) / ((sPred2 * sObs2)^0.25)
  Cb <- ((v + 1 / v + u^2)/2)^-1
  rCb <- r * Cb
  return(rCb)
}
```

There are several implementations for computing ρ_c with associated confidence intervals and p-value, and the reader can find examples in the `DescTools` package with the `CCC` function or in the `epiR` package with the `epi.ccc` function.

Ratio of performance to deviation (RPD) The ratio of performance to deviation (RPD) was proposed by Williams and Thompson (1978) as the ratio of standard error in prediction to the standard deviation. The objective of the RPD is to scale the error in prediction with the standard deviation of the property. It is widely used in the infrared spectroscopy literature as a way to assess the goodness of fit of infrared spectroscopy models.

The RPD is calculated as follows:

$$\mathrm{RPD} = \frac{\sqrt{\frac{1}{n-1}\sum_{i=1}^{n}(\mathrm{obs}_i - \overline{\mathrm{obs}})^2}}{\sqrt{\frac{1}{n}\sum_{i=1}^{n}(\mathrm{obs}_i - \mathrm{pred}_i)^2}}, \tag{9.7}$$

which is equivalent to `sd(obs)/RMSE(obs, pred)`. This metric and its systematic use have been criticized, in particular because the standard deviation of the soil property used to scale the error is misleading in the case of skewed or non-normal observations. The RPD is computed in R by:

```
RPD <- function(obs, pred){
  sdObs <- sd(obs)
  RMSE <- sqrt(mean((pred - obs)^2))
  rpd <- sdObs/RMSE
  return(rpd)
}
```

It can be seen that RPD is proportionally related to the coefficient of determination (or R2). R2 is based on variance, while RPD is based on standard error. If we assume a normal distribution, then `RPD = 1/sqrt(1-R2)`. We can test it (Fig. 9.1).

```
# create a sequence of number from 0 to 1
R2val <- seq(0, 1, by = 0.02)
RPDval = 1/sqrt(1 - R2val)

# plot the R2 and RPD
plot(R2val, RPDval, type = "l")
```

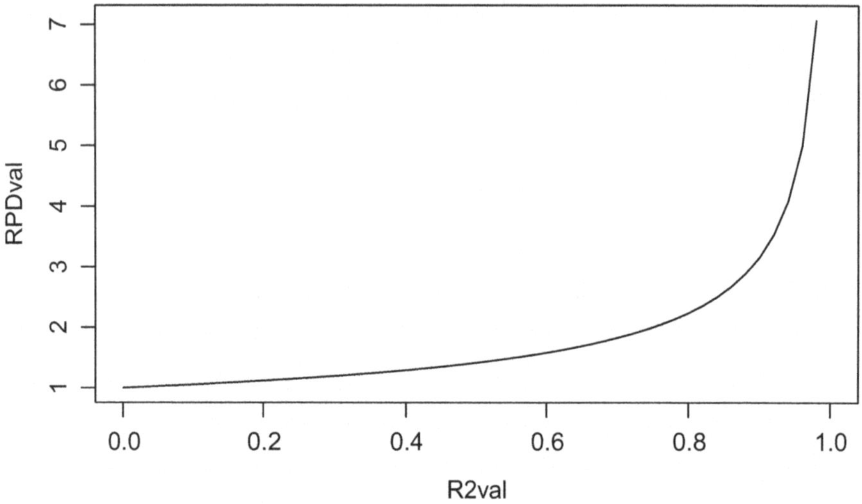

Fig. 9.1 Values of the R2 (x-axis) against those of the RPD (y-axis) computed on the validation dataset

Note that when R2 > 0.8, the RPD rapidly increases for larger R2 values. This suggests that a small increase in accuracy could be interpreted as a large boost in RPD.

Ratio of performance to inter-quartile distance (RPIQ) The ratio of performance to inter-quartile distance (RPIQ) has been proposed by Bellon-Maurel et al. (2010) to account for possibly non-normal distribution of the observations. The RPIQ is similar to the RPD, but it uses the inter-quartile range to represent the spread of the observations. It is therefore not sensitive to the statistical distribution of the observations. The values obtained by the RPIQ can be interpreted the same way than the RPD. The RPIQ is given by:

$$\text{RPIQ} = \frac{(Q_3(\text{obs}) - Q_1(\text{obs}))}{\sqrt{\frac{1}{n}\sum_{i=1}^{n}(\text{obs}_i - \text{pred}_i)^2}}, \tag{9.8}$$

where Q_1(obs) and Q_3(obs) are the first (25%) third (75%) quantiles of the observations (Q_3(obs) − Q_1(obs) is the inter-quartile distance) and the denominator is the RMSE. In R, it can be computed as follows:

```
RPIQ <- function(obs, pred){
  q25 <- as.numeric(quantile(obs)[2])
  q75 <- as.numeric(quantile(obs)[4])
  iqDist <- q75 - q25
  RMSE <- sqrt(mean((pred - obs)^2))
  rpiq <- iqDist/RMSE
  return(rpiq)
}
```

From the literature, it appears that various arbitrary limits of RPD were set to characterize a good model performance. For example, in agricultural products, it was quoted by Batten (1998) that RPD values greater than 3 are useful for screening, values greater than 5 can be used for quality control and values greater than 8 can be used for any application. In soil science, the paper by Chang et al. (2001) made other three categories: Category A, RDP > 2.0; Category B, RDP 1.4–2.0; and Category C, RDP < 1.4. This was interpreted by other authors as the reference standard in model performance: excellent if RPD > 2 and non-reliable models when RPD < 1.4. Other authors also have slightly modified this to justify the quality of prediction.

A soil scientist would say their model is excellent as it has RPD > 2, while a plant scientist would disagree as an excellent model should have RPD > 3. Limitations of RPD are described in Minasny and McBratney (2013), and the myth of RPD as a single measure of accuracy is summarized in the article of Esbensen et al. (2014): 'It is a myth that the RPD statistics furthers an objective, across-model, comparative, unambiguous prediction validation figure-of-merit'.

We warn against using an arbitrary classification system to justify the model performance, as RPD and R2 and other measures can be easily affected by the distribution of the data. We can illustrate how these accuracy measures are sensitive to the data distribution. Consider a set of random numbers (Fig. 9.2):

```
# set the seed for repeatability
set.seed(1)

# generate 100 numbers from an uniform distribution between 0 and 1
x = runif(100)
y = runif(100)

# plot the numbers
plot(x, y)

# add a 1:1 line to the plot
abline(a = 0, b = 1)
```

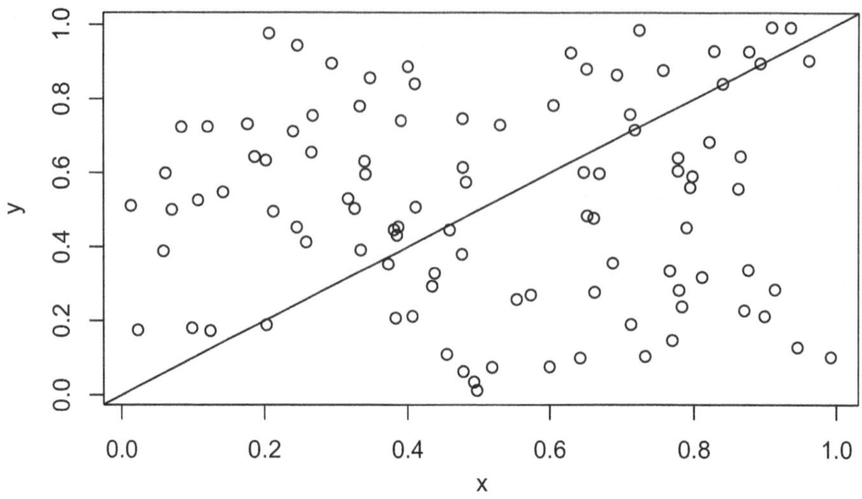

Fig. 9.2 Set of 100 randomly generated values between 0 and 1

Obviously the accuracy measure should say there is no clear relationship between the two random variables. We use the `eval` function from the book-associated `soilspec` package for this.

```
# evaluate the results
soilspec::eval(x, y, obj = "quant")
```

```
##    ME RMSE r2 R2 rhoC  RPD RPIQ
## 1   0 0.38  0 -1 0.02 0.71 1.18
```

Let us now see the effect of adding two extreme observations to the data (Fig. 9.3).

```
# generate two numbers from an uniform distribution between 8 and 10
x1 = runif(2, min = 8, max = 10)
y1 = runif(2, min = 8, max = 10)

# add the two new numbers to the vector of existing numbers
x = c(x, x1)
y = c(y, y1)

# plot the numbers
plot(x, y)

# add a 1:1 line to the plot
abline(a = 0, b = 1)
```

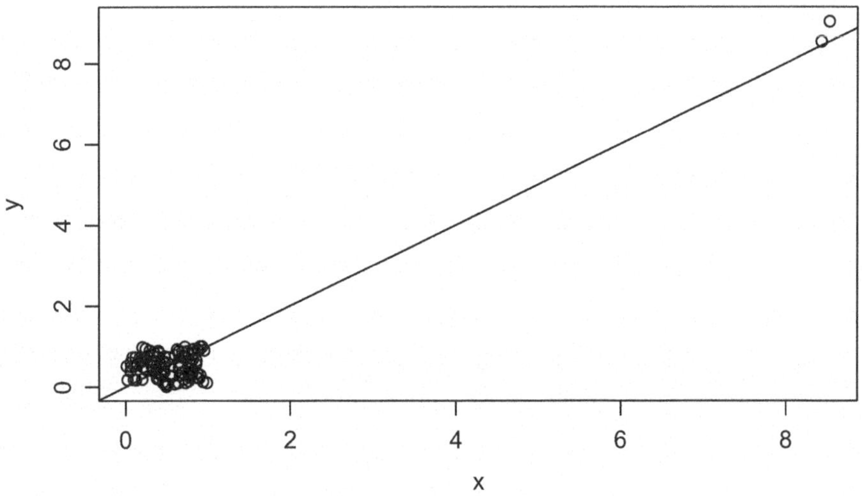

Fig. 9.3 Set of 100 randomly generated values between 0 and 1 and 2 values between 8 and 10

```
# evaluate the results
soilspec::eval(x, y, obj = "quant")
```

```
##      ME RMSE r2   R2 rhoC  RPD RPIQ
## 1 0.01 0.38  0 0.89 0.95 3.04 1.19
```

It is now visible that we can achieve RPD > 3, $\rho_c > 0.9$ and R2 = 0.9 indicating we have created an excellent model. The RPIQ seems not being much affected by outliers.

For categorical variables, the overall accuracy The overall accuracy (OA) measures the fraction of predictions that are correctly classified. Its optimal value is 1 (all if the predicted class equal the observed class) and falls to 0 if none of the predicted classes equal the observed classes. It is formally calculated as follows:

$$OA = \frac{\text{Number of correct prediction}}{\text{Total number of prediction}}, \tag{9.9}$$

where the numerator is an indicator having 1 if the predicted class equals the observed class and 0 otherwise and the denominator is the validation sample size. In R it is computed as follows:

```
OA <- function(obs, pred){
  # create a confusion matrix between observed and predicted classes
  cm = as.matrix(table(obs = obs, pred = pred))
  n <- length(obs)
  diag = diag(cm)
  OA <- sum(diag) / n
  return(OA)
}
```

As OA only calculates the overall accuracy, if the data is imbalanced. i.e. one class dominates over the others, a predictive model will only predict the dominant class and could provide high accuracy. To solve this problem, Cohen's kappa statistic is used.

For categorical variables, Cohen's kappa statistic Cohen's kappa statistic (Cohen 1960) is another measure of the agreement between predicted and observed classes. It is often referred to as the comparison of the overall accuracy to the expected random chance accuracy. Cohen's kappa is defined as the difference between the overall accuracy and the random chance accuracy divided by 1 minus the random chance accuracy. The statistic can be negative, but is more often comprised between 0 and 1, where 1 shows a perfect agreement between the predicted and observed classes and 0 no more agreement than what is expected by chance. It is derived as follows:

$$\text{Cohen's kappa} = \frac{\text{OA} - p_e}{1 - p_e}, \tag{9.10}$$

where OA is the overall accuracy derived previously and p_e is the expected probability of chance agreement. In other words, p_e is the expected random chance accuracy. In R Cohen's kappa statistic is computed as follows:

```
kappa <- function(obs, pred){
  # create a confusion matrix between observed and predicted classes
  cm = as.matrix(table(obs = obs, pred = pred))
  # number of observations per class
  rowsums = apply(cm, 1, sum)
  # number of predictions per class
  colsums = apply(cm, 2, sum)
  n <- length(obs)
  diag = diag(cm)
  accuracy <- sum(diag) / n
  p = rowsums / n # distribution of points over the actual classes
  q = colsums / n # distribution of points over the predicted classes
  expAccuracy = sum(p*q)

  kappa = (accuracy - expAccuracy) / (1 - expAccuracy)
  return(kappa)
}
```

9.2 Models for Quantitative Variables

A general problem in spectroscopy data is the large number of predictors, i.e. the number of spectral bands. The machine learning literature will describe this as a 'large p and small n' problem. In addition, the spectral bands are highly correlated.

One way of handling data with a high number of predictor variables such as in infrared spectroscopy is variable reduction. Another way is to select only relevant variables to use in the model (variable selection). Principal components and partial least squares regression (PLSR) methods are routinely used in chemometrics for variable reduction. Both models boil down to linear regression and principal component analysis. The PLSR model is particularly useful for prediction purposes. We will first explore these linear models and then continue with machine learning with cubist and random forest models.

In this section, we use the raw spectra described in Chap. 3 (Fig. 9.4). The steps for pre-processing are explained in the previous chapters.

```
# load the required packages
require(prospectr)
require(soilspec)

# load the data
data("datsoilspc")

# convert reflectance to absorbance
spectraA <- log(1/datsoilspc$spc)

# embed the soil property and the spectra in one single table
datsoilspc$spcA <- spectraA

# apply some smoothing to the spectra
oldWavs <- as.numeric(colnames(datsoilspc$spcA))
newWavs <- seq(min(oldWavs), max(oldWavs), by = 5)
datsoilspc$spcARs <- prospectr::resample(datsoilspc$spcA,
                                         wav = oldWavs,
                                         new.wav = newWavs,
                                         interpol = "linear")

# apply a standard normal variate transformation for baseline correction
datsoilspc$spcASnv <- standardNormalVariate(datsoilspc$spcARs)

# apply a moving average window to the standard normal variate spectra
datsoilspc$spcAMovav <- movav(datsoilspc$spcASnv, w = 11)

# convert the column names from integer to numeric
wavs <- as.numeric(colnames(datsoilspc$spcAMovav))

# plot first spectrum
matplot(x = wavs, y = t(datsoilspc$spcAMovav),
        xlab = "Wavelength /nm",
        ylab = "Absorbance",
        type = "l",
        lty = 1,
        col = rgb(red = 0.5, green = 0.5, blue = 0.5, alpha = 0.3))
```

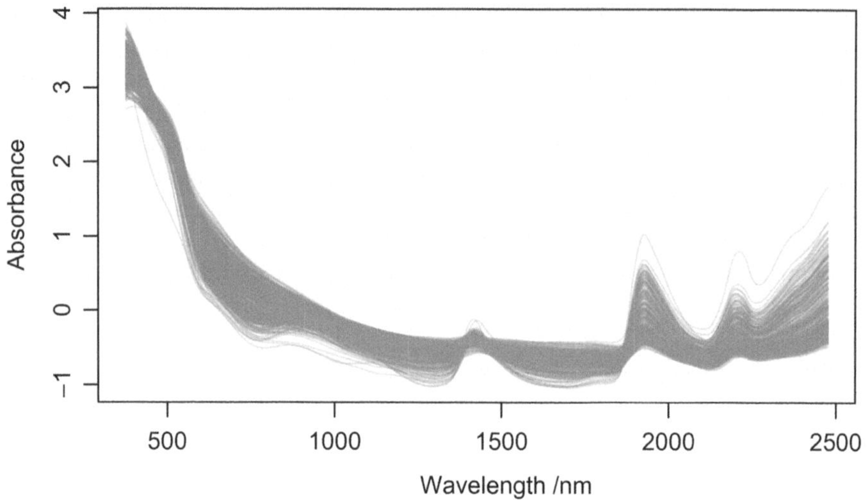

Fig. 9.4 Pre-processed absorbance spectra from the `datsoilspc` dataset provided in the book package `soilspec`

For the example, we further separate the data into calibration (75%) and validation (25%) by splitting randomly the dataset (Fig. 9.5).

```
# set the seed
set.seed(19101991)

# id of the rows to be used for calibration
calId <- sample(1:nrow(datsoilspc), size = round(0.75*nrow(datsoilspc)))

# separate the dataset into calibration and validation
datC <- datsoilspc[calId,]
datV <- datsoilspc[-calId,]

# plot the value of the Total Carbon content for both calibration and validation
par(mfrow=c(1,2))

# calibration
hist(datC$TotalCarbon,
     main = "",
     xlab = "Total carbon")

# validation
hist(datV$TotalCarbon,
     main = "",
     xlab = "Total carbon")
```

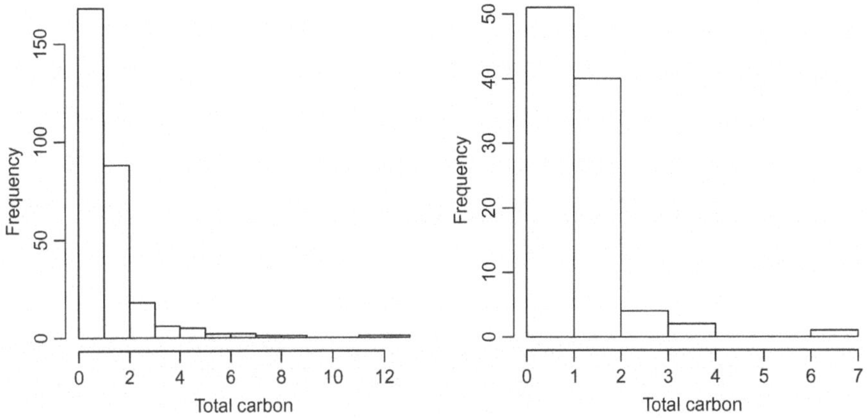

Fig. 9.5 Histograms of the soil total carbon content from the Geeves et al. (1994) dataset provided in the book package `soilspec` for both calibration (left) and validation (right)

The observation of the total carbon content is slightly skewed to the left. In a real-world case study, we might want to test whether a transformation makes the carbon values look normally distributed. The slight skewness is not critical in this case study.

9.2.1 Principal Component Regression

Principal component regression (PCR) boils down to principal component analysis (Sect. 6.2) and multiple linear regression by taking the principal component of the spectra and by building a linear regression on the component scores. Recall the matrix of scores $\mathbf{T}$ obtained by PCA of the matrix of spectral variables $\mathbf{X}$. In PCR, a linear regression model is fitted between the scores of the PC of $\mathbf{X}$ and the soil property (response variable) $\mathbf{y}$ (if only one soil property is predicted) or $\mathbf{Y}$ of size $n \times c$ where c is the number of soil properties of interest ($c = 1$ for a single response). The PCR model takes the following form:

$$\mathbf{Y} = \mathbf{T}\boldsymbol{\beta} + \mathbf{E}, \tag{9.11}$$

where $\boldsymbol{\beta}$ denotes the vector of regression coefficients, found by solving $\boldsymbol{\beta} = (\mathbf{X}^T\mathbf{X})^{-1}\mathbf{X}^T\mathbf{Y}$, and $\mathbf{E}$ is a matrix of residuals, the same size than $\mathbf{Y}$.

The user must then decide how many components to use in the model. Usually, the optimal number of components is defined by cross-validation. The cross-validation can be done by using the `pls` package, but for illustration we will use the `princomp` and `lm` functions.

```
# first we perform a PCA on the spectra
pcspectra <- prcomp(datC$spcAMovav,
                    center = TRUE, scale = TRUE)

# calculate the percent of the variances explained by the PC
v <- pcspectra$sdev*pcspectra$sdev

# percentage of cumulative variances
cumv <- 100*cumsum(v)/sum(v)

# plot cumulative percentage of variances explained by the PCs
plot(cumv[1:20],
     type = "b",
     xlab = "PC",
     ylab = "% Cumulative variance")
```

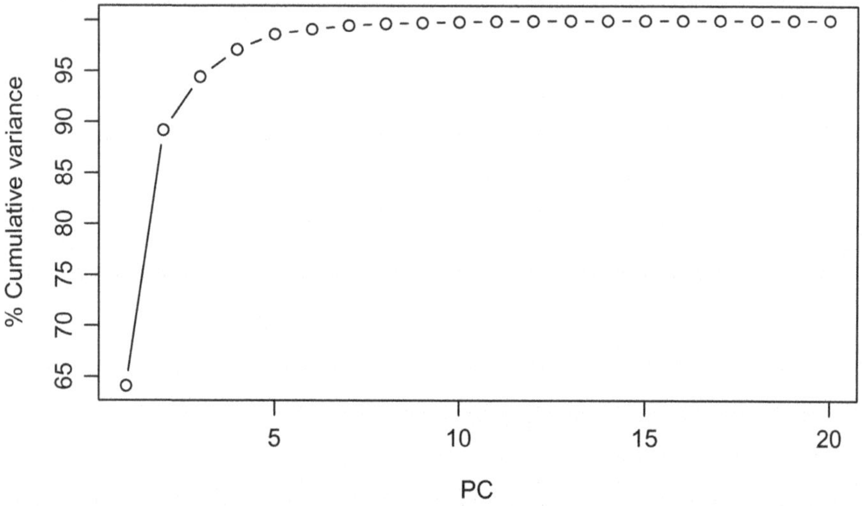

Fig. 9.6 Percentage of variance explained by the principal components against number of components

Figure 9.6 shows that around nine components capture more than 99% of the variation.

One can now fit a simple PC regression. First we specify the number of principal components to use in the model and then form the PC scores as the independent variables and soil property as the dependent variable in a linear model.

```
# specify number of components
npc <- 9

# select PC scores
sdata <- as.data.frame(pcspectra$x[,1:npc])

# fit a linear model Total C = PC1 + PC2 + ...
soilCPcrModel <- lm(datC$TotalCarbon ~ ., data = sdata)
```

```
# obtain a summary of the fit
summary(soilCPcrModel)
```

```
##
## Call:
## lm(formula = datC$TotalCarbon ~ ., data = sdata)
##
## Residuals:
##     Min      1Q  Median      3Q     Max
## -2.2051 -0.4448 -0.0744  0.3529  4.5404
##
## Coefficients:
##              Estimate Std. Error t value Pr(>|t|)
## (Intercept)  1.214198   0.046704  25.998  < 2e-16 ***
## PC1          0.027437   0.002848   9.635  < 2e-16 ***
## PC2          0.059561   0.004554  13.080  < 2e-16 ***
## PC3          0.001337   0.009975   0.134  0.89348
## PC4          0.157447   0.013955  11.282  < 2e-16 ***
## PC5          0.303688   0.018495  16.420  < 2e-16 ***
## PC6          0.014171   0.032577   0.435  0.66389
## PC7          0.115872   0.036931   3.138  0.00188 **
## PC8          0.400682   0.053989   7.422 1.36e-12 ***
## PC9         -0.043479   0.076195  -0.571  0.56871
## ---
## Signif. codes:  0 '***' 0.001 '**' 0.01 '*' 0.05 '.' 0.1 ' ' 1
##
## Residual standard error: 0.7994 on 283 degrees of freedom
## Multiple R-squared:  0.7196, Adjusted R-squared:  0.7107
## F-statistic:  80.7 on 9 and 283 DF,  p-value: < 2.2e-16
```

We can now assess the goodness of the fit by plotting the observed versus predicted values of the total carbon, for both the calibration and validation datasets (Fig. 9.7).

```
# predict on the calibration dataset
soilCPcrPred <- predict(soilCPcrModel, sdata)

# predict on the validation dataset
pcspectraV <- predict(pcspectra, datV$spcAMovav)
sdataNew <- as.data.frame(pcspectraV[, 1:npc])
soilVPcrPred <- predict(soilCPcrModel, sdataNew)

par(mfrow = c(1, 2))

# plot calibration
plot(datC$TotalCarbon, soilCPcrPred,
     xlab = "Observed",
     ylab = "Predicted",
     xlim = c(0, 12),
     ylim = c(0, 12),
     pch = 16)
abline(0, 1)

# plot validation
plot(datV$TotalCarbon, soilVPcrPred,
```

```
       xlab = "Observed",
       ylab = "Predicted",
       xlim = c(0, 12),
       ylim = c(0, 12),
       pch = 16)
abline(0, 1)
```

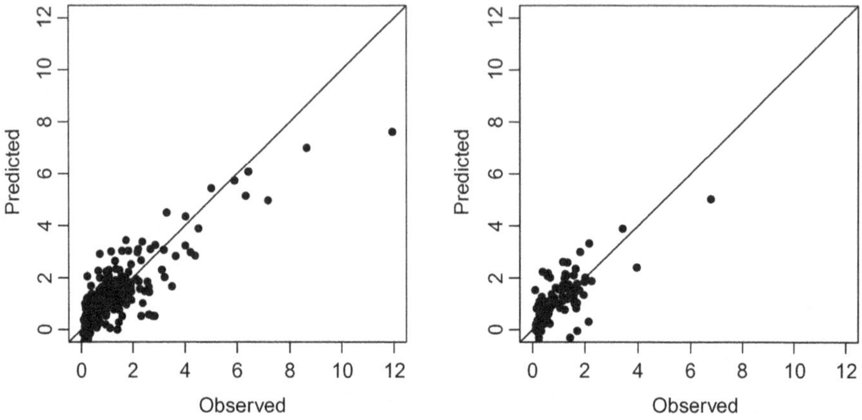

Fig. 9.7 Scatterplot of observed versus predicted value of the total carbon. The predictions are made by principal component regression

Derive some accuracy measures, for the example using the `eval` function and specifying the argument `obj = "quant"` for continuous variables:

```
# accuracy measures for calibration
soilspec::eval(datC$TotalCarbon, soilCPcrPred, obj = "quant")
```

```
##    ME RMSE   r2   R2 rhoC  RPD RPIQ
## 1   0 0.79 0.57 0.72 0.84 1.89 1.46
```

and for validation.

```
# accuracy measures for validation
soilspec::eval(datV$TotalCarbon, soilVPcrPred, obj = "quant")
```

```
##      ME RMSE   r2   R2 rhoC  RPD RPIQ
## 1 0.08 0.74 0.42 0.36  0.7 1.26 1.63
```

9.2.2 *Partial Least Squares Regression*

Partial least squares regression (PLSR) is a technique that attempts to combine PCA and multiple regression (Wold et al. 2001). It aims to predict a set of dependent variables (soil properties) by extracting from the spectra a set of 'orthogonal' factors (or latent variables) which give the best prediction. The components in partial least squares are determined by the predictor variables **X** (as in PCR) but also by the response variable(s) (**y** or **Y** if multiple responses). The PLSR model takes the following form:

$$\mathbf{X} = \mathbf{T}\mathbf{P}^T + \mathbf{E}, \tag{9.12}$$

$$\mathbf{Y} = \mathbf{U}\mathbf{Q}^T + \mathbf{F}, \tag{9.13}$$

where **X**, **T** and **P** are defined previously. Both **T** and **U** are score matrices of size $n \times p$ for **X** or **Y**, respectively, and **P** (size $d \times b$) and **Q** (size $d \times c$) are loading matrices for **X** or **Y**. Finally, **E** is the matrix of residuals of size $n \times b$ for **X** and **F** is the matrix of residuals of size $n \times c$ for **Y**. A regression between **X** and **Y** is obtained by:

$$\mathbf{U} = \mathbf{T}\boldsymbol{\beta}, \tag{9.14}$$

where $\boldsymbol{\beta}$ is the vector of regression coefficients of the linear model. Substituting this relationship from the original model, predictions are obtained by:

$$\mathbf{Y} = \mathbf{U}\mathbf{Q}^T = \mathbf{T}\boldsymbol{\beta}\mathbf{Q}^T. \tag{9.15}$$

This is implemented in R with the `plsr` function from the `pls` package, made by Wehrens and Mevik (2007). As for the PCA and PCR, we must choose the optimal number of principal components (Fig. 9.8).

```
# load required package
require(pls)

# maximum number of components in the PLS model
maxc <- 30

# generate a PLS model based on calibration data
soilCPlsModel <- plsr(TotalCarbon ~ spcAMovav,
                      data = datC,
                      method = "oscorespls",
                      ncomp = maxc,
                      validation = "CV")

# this is the plsr function, using cross validation to evaluate the RMSEP
# as a function of number of components from one until maxc
plot(soilCPlsModel, "val",
     main = " ",
     xlab = "Number of components")
```

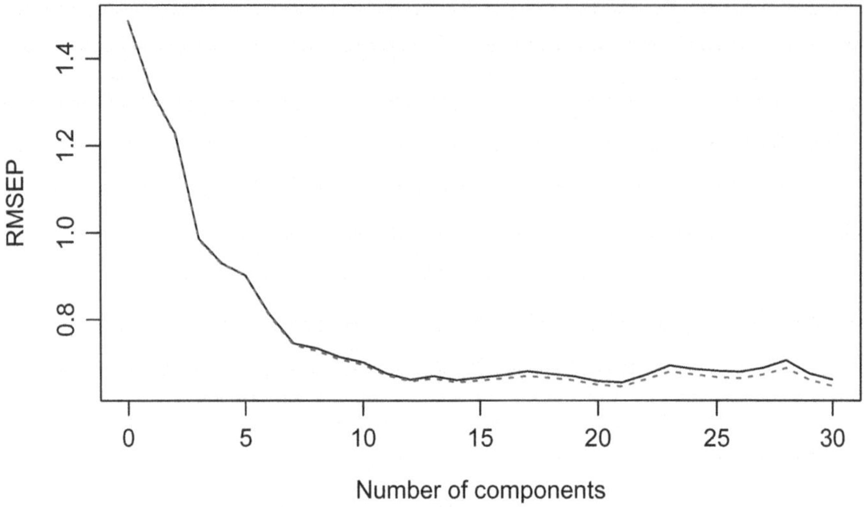

Fig. 9.8 Root mean square error of the prediction (RMSEP, black line) and bias-adjusted RMSEP (red dashed line) obtained by cross-validation against the number of components used in the PLSR model

More information on the RMSEP and bias-adjusted RMSEP values are obtained in the vignette of the `pls` package. Note that we use `method = oscorespls`, which is not the default of the `plsr` function. It is discussed later in this section. This figure from a 10-fold cross-validation on the calibration data shows that 14 components seem to produce a minimal RMSEP. So we use this number.

```
# number of components to use
nc <- 14
```

It is also possible to plot directly the predicted and observed values of the soil properties. The predictions are made using the fitted `pls_model` using `nc` principal components (Fig. 9.9).

```
# plot of cross-validated predictions
plot(soilCPlsModel,
     ncomp = nc,
     main = " ",
     xlab = "Observed",
     ylab = "Predicted")
```

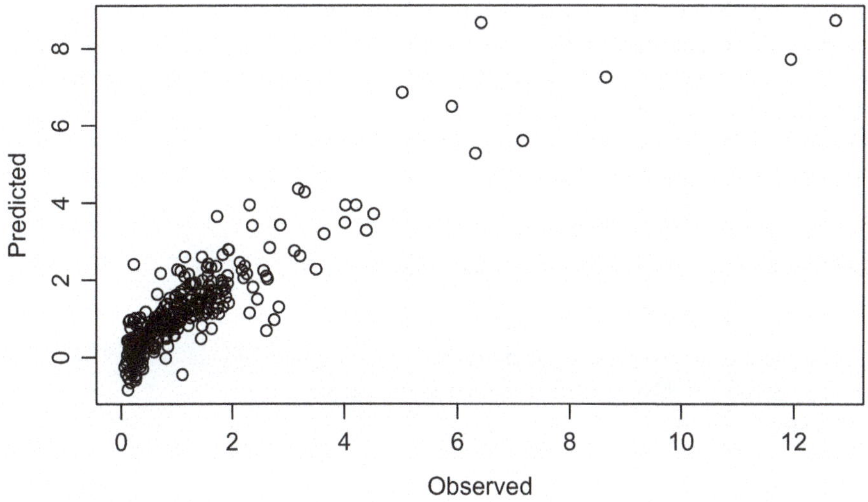

Fig. 9.9 Scatterplot of observed against predicted values of the total carbon. The predictions are made by partial least squares regression

This method plots observed against predicted based on the cross-validated predictions using `nc` number of components. Additional figures can be derived from the `soilCPlsModel` object. Note that the values in the x-axis are the indices of the wavelength (Fig. 9.10).

```
# the three first loadings
plot(soilCPlsModel,
     "loadings",
     comps = 1:3,
     xlab = "Index of the wavelength",
     ylab = "Loading value")
```

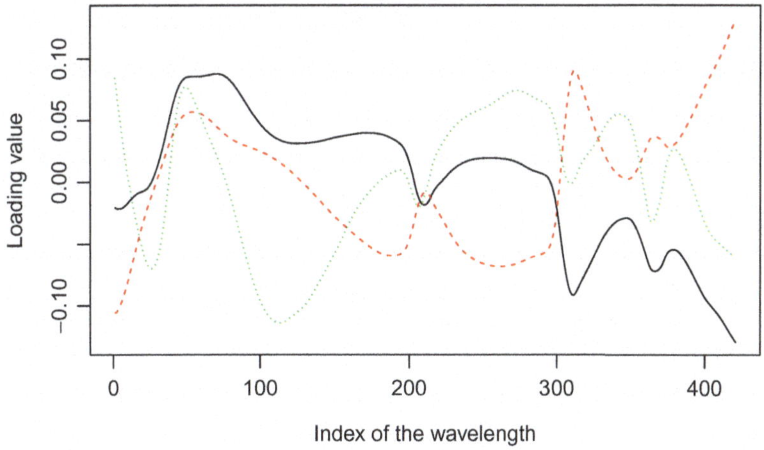

Fig. 9.10 First three loadings of the principal components of the spectra

Figure 9.10 shows the three first loadings of the PLS components of the spectra. It gives an idea which wavelengths have the most influence on each of the components, which can be used for spectral interpretation. In addition to the loadings of the PC, in PLSR the regression coefficients can be plotted. We do not use the default `plot` function of the `plsr` package and select instead manually the regression coefficient for the case of `nc = 14` (Fig. 9.11).

```
# plot the coefficient
plot(wavs, soilCPlsModel$coefficients[,1,nc],
     main = " ",
     type = "l",
     xlab = "Wavelength /nm",
     ylab = "Regression coefficient")
abline(h = 0)
```

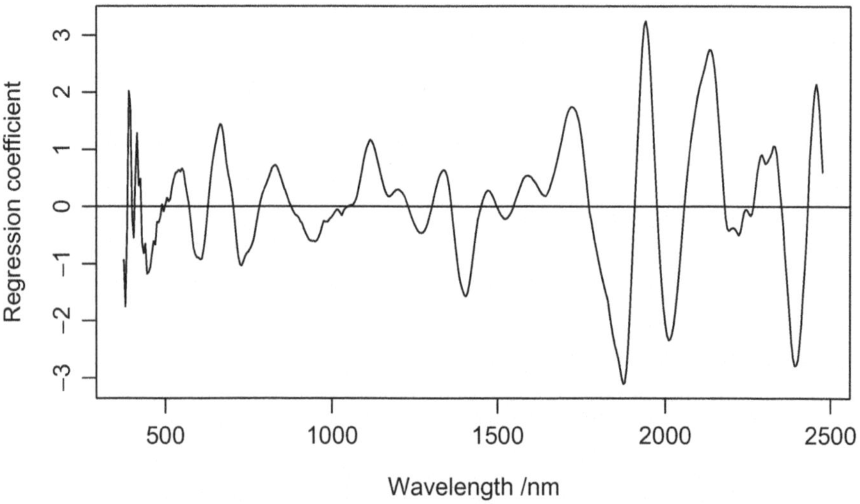

Fig. 9.11 Standardized regression coefficient of the PLSR model for predicting total carbon

This gives a more meaningful interpretation, the standardized coefficients of the regression. Wavelengths with a large (positive or negative) value of the regression coefficient are more influential in the prediction.

Alternative to the regression coefficients and the principal component loadings, one can use the variable importance on projection (VIP) proposed by Wold et al. (1993). It measures the importance of each wavelength in the projection of the components. It is calculated as a weighted sum of the squares of the PLS weights

(Ng et al. 2019). Influential variables have a variable importance on projection value greater than 1. As there is no direct implementation in the `pls` package, we provide the code below. For more information, the reader is redirected to the article of Wold et al. (1993). Note that the VIP calculation is valid only for the orthogonal score algorithm (`method = "oscorespls"` in the `pls` package) and for a single response PLSR model. By default, the `plsr` function in R uses the `kernelpls` algorithm.

```
# take the loadings, loading weights and scores
W <- soilCPlsModel$loading.weights
Q <- soilCPlsModel$Yloadings
TT <- soilCPlsModel$scores

# compute the variable importance, see Wold et al., (1993)
Q2 <- as.numeric(Q) * as.numeric(Q)
Q2TT <- Q2[1:nc] * diag(crossprod(TT))[1:nc]
WW <- W * W/apply(W, 2, function(x) sum(x * x))
vip <- sqrt(length(wavs) * apply(sweep(WW[, 1:nc], 2, Q2TT, "*"),
                                 1, sum)/sum(Q2TT))

# display the variable importance
plot(wavs, vip,
     xlab = "Wavelength /nm",
     ylab = "Importance",
     type = "l",
     lty = 1,
     col = rgb(red = 0.5, green = 0.5, blue = 0.5, alpha = 1))
abline(h = 1)
```

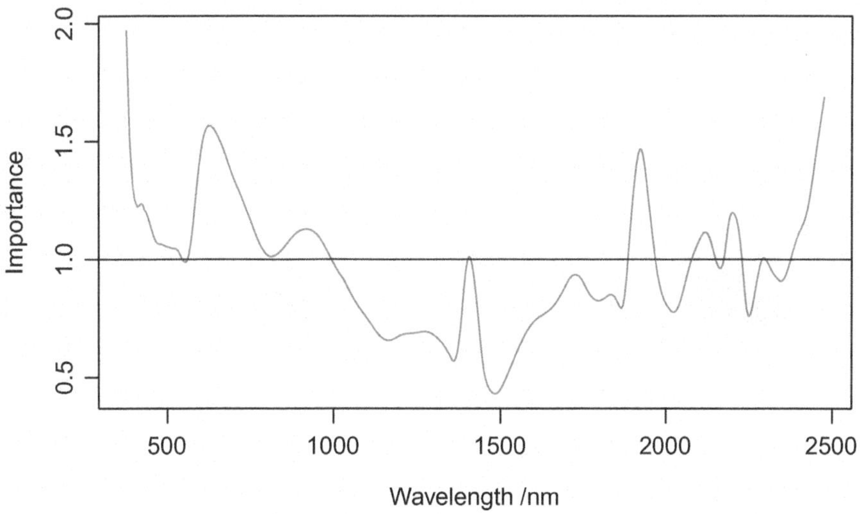

Fig. 9.12 Importance of each wavelength on projection of the principal components

Figure 9.12 shows that the important wavelengths in the projection of the PLS components to predict the total soil carbon are similar to the wavelengths found important in the previous figures using the standardized regression coefficient or the first three principal component loadings.

Now that we have created the PLS model, we can use the model to predict using the spectra on the calibration and validation datasets (Fig. 9.13).

```
# predict on the calibration dataset
soilCPlsPred <- predict(soilCPlsModel, ncomp = nc, newdata = datC$spcAMovav)

# predict on the validation dataset
soilVplsPred <- predict(soilCPlsModel, ncomp = nc, newdata = datV$spcAMovav)

par(mfrow = c(1, 2))

# plot calibration
plot(datC$TotalCarbon, soilCPlsPred,
     xlab = "Observed",
     ylab = "Predicted",
     xlim = c(0, 12),
     ylim = c(0, 12),
     pch = 16)
abline(0, 1)

# plot validation
plot(datV$TotalCarbon, soilVplsPred,
     xlab = "Observed",
     ylab = "Predicted",
     xlim = c(0, 12),
     ylim = c(0, 12),
     pch = 16)
abline(0, 1)
```

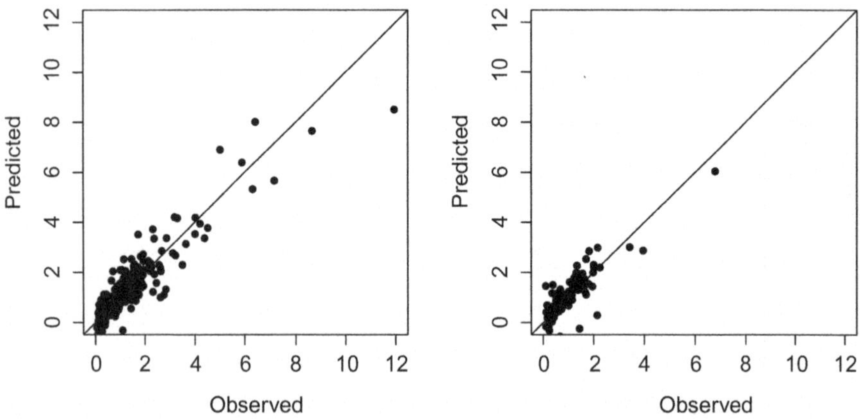

Fig. 9.13 Scatterplot of observed against predicted values of the total carbon. Predictions are made by a PLSR model. The left-hand side plot shows the observed against predicted values for the calibration dataset, while the right-hand side plot shows the observations against predictions for the validation dataset

Derive some accuracy measures, for the example using the `eval` function and specifying the argument `obj = "quant"` for continuous variables:

```
# accuracy measures for calibration
soilspec::eval(datC$TotalCarbon, soilCPlsPred, obj = "quant")
```

```
##    ME RMSE  r2   R2 rhoC  RPD RPIQ
## 1   0 0.56 0.8 0.86 0.92 2.63 2.04
```

and for validation.

```
# accuracy measures for validation
soilspec::eval(datV$TotalCarbon, soilVplsPred, obj = "quant")
```

```
##      ME RMSE   r2   R2 rhoC  RPD RPIQ
## 1 0.02 0.52 0.64 0.69 0.85 1.81 2.34
```

Bagging PLSR

One way of 'strengthening' the PLSR prediction is to generate multiple models and average the prediction (making an ensemble model). Bootstrap aggregating or bagging (Breiman 1996) manipulates the calibration data to generate different models. The bootstrap is a general statistical method used to assess the accuracy of a prediction by sampling the calibration data with replacement. Suppose the calibration data of size n is composed of predictors and response; we randomly generate B datasets based on the calibration data by sampling with replacement. For each of the bootstrap datasets, we fit a PLSR model. The bagging estimate is calculated as the average of all the model predictions.

Therefore, it combines the outputs of many models to produce a powerful 'committee', which is useful when dealing with data with high variation, as each realization will produce a model that fits a particular set of the data which may differ from other realizations. The important element of bagging is that by perturbing the calibration, it can cause significant changes in the predictor. The aggregated predictor averages the prediction over a collection of bootstrap samples, therefore reducing the variance of prediction. The accuracy of the prediction is increased when the prediction method is unstable, i.e. small changes in the calibration data used in bootstrap can result in large changes in the resulting predictor. It was used by McBratney et al. (2006) for soil prediction and quantifying its uncertainty. The improvement over a single PLSR could be small, but it may be more robust against noise in the spectra, and it is also possible to obtain uncertainty intervals of the prediction (Mevik et al. 2004).

A function called `fitBagPlsr` was created for this.

```
fitBagPlsr <- function (soilv, spec, nbag, maxc){
  nc <- maxc
  n <- length(soilv)
  vPls <- vector(nbag, mode = "list")
  calRmse <- matrix(0, nrow = nbag, ncol = 1)
  oobRmse <- matrix(0, nrow = nbag, ncol = 1)

  for (ibag in 1:nbag){
    # take a bootstrap sample with replacement
    s <- sample.int(n, replace = TRUE)
```

```
    #build a plsr model
    vPls[[ibag]] <- plsr(soilv[s] ~ spec[s, ], maxc)

    # compute calibration RMSE
    predV <- predict(vPls[[ibag]], ncomp = nc, newdata = spec[s,])
    err2 <- (soilv[s] - predV)^2
    calRmse[ibag] <- sqrt(mean(err2))

    # compute out-of-bag RMSE
    predC <- predict(vPls[[ibag]], ncomp = nc, newdata = spec[-s,])
    err2 <- (soilv[-s] - predC)^2
    oobRmse[ibag] <- sqrt(mean(err2))
  }

  # average the results
  avCalRmse <- mean(calRmse)
  avOobRmse <- mean(oobRmse)

  # return the results
  list(modelBpls = vPls, oobRmse = avOobRmse, calRmse = avCalRmse)
}
```

Based on our previous single PLS model, we use bagging to generate 50 bootstrapped models.

```
# number of bootstrap
nbag <- 50

# maximum number of components
maxc <- 14

# number of components used in the PLSR model, set to equal to maxc
nc <- maxc

# make the bootstrap
bagPlsr <- fitBagPlsr(datC$TotalCarbon, datC$spcAMovav,
                      nbag,
                      maxc)
```

The output of the function contains `model` which is the bootstrapped PLSR models, `oobRmse` which is the mean out-of-bag RMSE and `calRmse` which is the mean calibration RMSE.

Out of bag is the internal validation used in bootstrap. At each bootstrap, a sample of size *n* of the original data was sampled with replacement. That means that for each bootstrap about one-third of the data are not used in the calibration. This oob (out-of-bag) data are used to estimate the error of the model.

```
# average RMSE from oob estimates
bagPlsr$oobRmse
```

```
## [1] 0.7093668
```

```
# average RMSE from calibration
bagPlsr$calRmse
```

```
## [1] 0.5187187
```

Now that we have created the bagged PLS model, we can use the model to predict using the spectra on the calibration and validation datasets using the following function.

```
predictBagPlsr <- function(modelBpls, newspec, nbag, nc){

  n <- nrow(newspec)
  predV <- matrix(0, nrow = n, ncol = nbag)

  for (ibag in 1:nbag) {
    predV[,ibag] <- predict(modelBpls[[ibag]], ncomp = nc, newdata = newspec)
  }
  predAve <- apply(predV, 1, mean)
  predStd <- apply(predV, 1, sd)

  return(list(bagPred = predV, predAve = predAve, predStd = predStd))
}
```

In the function `predictBagPlsr`, the argument `modelBpls` is the bagged PLSR model, `newspec` is the spectra of the validation set, `nbag` is the number of bootstrap samples, and `nc` is the number of PCs used in PLSR. The function returns `bagPred` that is the bagged predicted values, `predAve` that is the mean of the predictions and `predStd` that is the standard deviation of the predictions (Fig. 9.14).

```
# predict on the calibration dataset
soilCBagplsPred <- predictBagPlsr(bagPlsr$modelBpls, datC$spcAMovav,
                                  nbag,
                                  nc)

# predict on the validation datset
soilVBagplsPred <- predictBagPlsr(bagPlsr$modelBpls, datV$spcAMovav,
                                  nbag,
                                  nc)

par(mfrow = c(1, 2))

# plot calibration
plot(datC$TotalCarbon, soilCBagplsPred$predAve,
     xlab = "Observed",
     ylab = "Predicted",
     xlim = c(0, 12),
     ylim = c(0, 12),
     pch = 16)
abline(0, 1)

# plot validation
plot(datV$TotalCarbon, soilVBagplsPred$predAve,
     xlab = "Observed",
     ylab = "Predicted",
     xlim = c(0, 12),
     ylim = c(0, 12),
     pch = 16)
abline(0, 1)
```

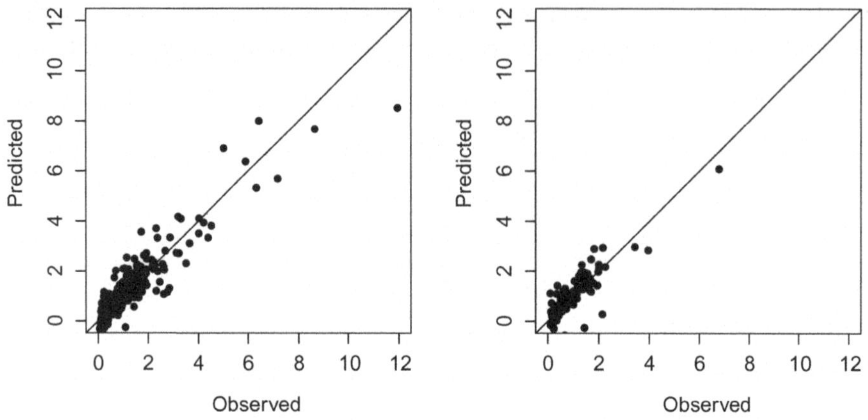

Fig. 9.14 Scatterplot of observed against predicted values of the total carbon. Predictions are made using a bagged partial least squares regression model

Derive some accuracy measures, for the example using the `eval` function and specifying the argument `obj = "quant"` for continuous variables:

```
# accuracy measures for calibration
soilspec::eval(datC$TotalCarbon, soilCBagplsPred$predAve, obj = "quant")
```

```
##    ME RMSE   r2   R2 rhoC  RPD RPIQ
## 1   0 0.56 0.79 0.86 0.92 2.64 2.04
```

and for validation.

```
# accuracy measures for validation
soilspec::eval(datV$TotalCarbon, soilVBagplsPred$predAve, obj = "quant")
```

```
##      ME RMSE   r2  R2 rhoC  RPD RPIQ
## 1 0.01  0.5 0.66 0.7 0.86 1.85 2.39
```

9.2.3 *Cubist*

Another way of handling large dimensional data is using variable selection techniques to find the best predictors. When the high dimensional data has been reduced to several components or important variables have been selected, they are used for prediction using either linear regression or data-mining tools. Regression trees, neural networks and support vector machines have been used for such predictions.

There are also data-mining tools which are designed to extract information on data containing large number of variables and large number of samples. This is potentially useful as the data reduction step need not be taken. Models that improve regression trees have been proposed, including random forest. While RF has been used successfully for prediction, the model form is complex, and interpretation can be difficult as no explicit formulae can be given.

Other forms of data-mining tools based on the idea of decision trees are the regression rules, rule-based regression or cubist model. This is in effect transforming regression into a classification problem, the model consists of a set of rules, and each rule consists of a linear model. The idea is similar to the regression tree algorithm; while regression trees have a value at each 'leaf', regression rules build a multivariate linear function. Regression rules are also analogous to piecewise linear functions.

The cubist model takes the form of:

```
Rule 1: [10 cases, mean -0.96, range -1.77 to 0.66, est err 0.27]
       if
    R880 <= 0.0144
    R1610 > -0.921
      then
    outcome = -4.06 + 1.27 * R540 + 1.61 * R1610
```

The model has several rules. Each rule has a 'condition' (reflectance at 880 nm <= 0.0144 & at 1610 nm > -0.921); if this condition is met by the data, then the prediction is the given linear function. The program also informs the statistics of each rule which refers to the range of values of the predicted and also the error of the model.

Cubist initially was a commercial regression-rules program, but now a public GNU code has been provided and ported in `Cubist` R package by Kuhn et al. (2012). The model of cubist is a set of comprehensible rules, where each rule has an associated linear model. Whenever a situation matches a rule's conditions, the associated model is used to calculate the predicted value. The first use of cubist for soil spectroscopy modelling was made by Minasny and McBratney (2008).

```
# load required package
require(Cubist)

# make a Cubist model on calibration dataset
soilCCubistModel <- cubist(x = datC$spcAMovav, y = datC$TotalCarbon)

# summary of the model
summary(soilCCubistModel)
```

```
##
## Call:
## cubist.default(x = datC$spcAMovav, y = datC$TotalCarbon)
##
##
## Cubist [Release 2.07 GPL Edition]  Tue Aug 25 15:51:17 2020
## ---------------------------------
##
##     Target attribute 'outcome'
##
## Read 293 cases (422 attributes) from undefined.data
##
## Model:
##
##   Rule 1: [106 cases, mean 0.347, range 0.06 to 1.94, est err 0.120]
##
##     if
##  1415 > -0.4015094
##     then
##  outcome = 4.663 - 447.07 850 + 1223.8 1410 - 1089.7 1400 + 326.8 860
##            - 240.31 2305 + 277.82 845 + 220.88 2310 - 493.2 1415
##            - 162.2 865 + 133.4 2165 + 458.8 1395 - 111.19 2170
##            + 150.9 2075 - 60 810 + 56.88 825 - 119.1 2100 - 36.37 625
```

```
##              + 35.28 630 - 55.7 2015 - 70.4 1815 + 65.8 1805 - 131.6 1455
##              + 32.19 2285 - 94.3 1405 + 22.26 800 + 95.1 1465 + 13.9 1940
##              + 39.6 1365 - 12.23 2380 + 54.4 1435 - 12.34 790 - 47.2 1430
##              + 9.65 2345 - 9.11 2350 - 9 2120 - 6 885 + 2.69 645
##              - 14.6 1445 + 5.7 2145 + 3.3 910
## etc... [shortened]
```

The summary of the model provides the full model. It also informs the rules in the model and the number of times (frequency) certain variables (wavelengths) are used as conditions and as predictors. We can plot these as an indicator of which wavelengths are useful in the model, which we will describe later.

The calibrated cubist model is then used to predict, on both the calibration and validation data (Fig. 9.15).

```
# predict on the calibration data
soilCCubistPredict <- predict(soilCCubistModel, datC$spcAMovav)

# predict on the calibration data
soilVCubistPredict <- predict(soilCCubistModel, datV$spcAMovav)

par(mfrow = c(1, 2))

# plot calibration
plot(datC$TotalCarbon, soilCCubistPredict,
     xlab = "Observed",
     ylab = "Predicted",
     xlim = c(0, 12),
     ylim = c(0, 12),
     pch = 16)
abline(0, 1)

# plot validation
plot(datV$TotalCarbon, soilVCubistPredict,
     xlab = "Observed",
     ylab = "Predicted",
     xlim = c(0, 12),
     ylim = c(0, 12),
     pch = 16)
abline(0, 1)
```

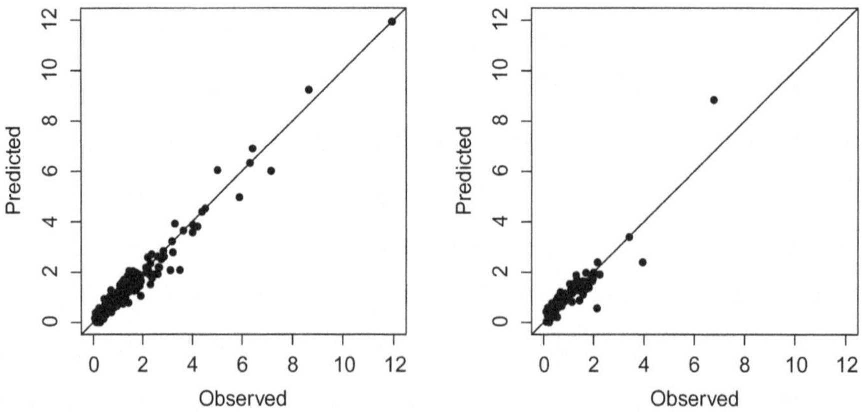

Fig. 9.15 Scatterplot of observed against predicted values of the total carbon. Predictions are made using a cubist model

Derive some accuracy measures, for the example using the `eval` function and specifying the argument `obj = "quant"` for continuous variables:

```
# accuracy measures for calibration
soilspec::eval(datC$TotalCarbon, soilCCubistPredict, obj = "quant")
```

```
##         ME RMSE   r2   R2 rhoC  RPD RPIQ
## 1 -0.02 0.27 0.91 0.97 0.98 5.46 4.23
```

and for validation.

```
# accuracy measure for validation
soilspec::eval(datV$TotalCarbon, soilVCubistPredict, obj = "quant")
```

```
##      ME RMSE  r2   R2 rhoC RPD RPIQ
## 1 0.03 0.39 0.8 0.82 0.92 2.4 3.11
```

Note that the validation statistics show that the calibrated cubist model actually gives us a worse prediction compared to PLS. From the model summary, we can also see that rules 4, 7 and 8 are only fitted to eight, five and seven observations. This could make the model overfit the data. So we need to simplify the model, by setting fewer rules. We re-run the model using two rules by adding the following options: `control = cubistControl(rules = 2)`.

```
# make a Cubist model on calibration dataset with 2 rules
soilCCubistModel2 <- cubist(x = datC$spcAMovav, y = datC$TotalCarbon,
                            control = cubistControl(rules = 2))

# summary of the model
summary(soilCCubistModel)
```

```
##
## Call:
## cubist.default(x = datC$spcAMovav, y = datC$TotalCarbon)
##
##
## Cubist [Release 2.07 GPL Edition]  Tue Aug 25 15:51:17 2020
## ---------------------------------
##
##     Target attribute `outcome'
##
## Read 293 cases (422 attributes) from undefined.data
##
## Model:
##
##   Rule 1: [106 cases, mean 0.347, range 0.06 to 1.94, est err 0.120]
##
##     if
##  1415 > -0.4015094
##     then
##  outcome = 4.663 - 447.07 850 + 1223.8 1410 - 1089.7 1400 + 326.8 860
##            - 240.31 2305 + 277.82 845 + 220.88 2310 - 493.2 1415
##            - 162.2 865 + 133.4 2165 + 458.8 1395 - 111.19 2170
##            + 150.9 2075 - 60 810 + 56.88 825 - 119.1 2100 - 36.37 625
##            + 35.28 630 - 55.7 2015 - 70.4 1815 + 65.8 1805 - 131.6 1455
##            + 32.19 2285 - 94.3 1405 + 22.26 800 + 95.1 1465 + 13.9 1940
##            + 39.6 1365 - 12.23 2380 + 54.4 1435 - 12.34 790 - 47.2 1430
##            + 9.65 2345 - 9.11 2350 - 9 2120 - 6 885 + 2.69 645
##            - 14.6 1445 + 5.7 2145 + 3.3 910
```

```
##
##   Rule 2: [155 cases, mean 1.263, range 0.31 to 3.5, est err 0.280]
## etc... [shortened]
```

The calibrated cubist model is then used to predict, on both the calibration and validation data (Fig. 9.16).

```
# predict on the calibration data
soilCCubist2Predict <- predict(soilCCubistModel2, datC$spcAMovav)

# predict on the calibration data
soilVCubist2Predict <- predict(soilCCubistModel2, datV$spcAMovav)

par(mfrow = c(1, 2))

# plot calibration
plot(datC$TotalCarbon, soilCCubist2Predict,
     xlab = "Observed",
     ylab = "Predicted",
     xlim = c(0, 12),
     ylim = c(0, 12),
     pch = 16)
abline(0, 1)

# plot validation
plot(datV$TotalCarbon, soilVCubist2Predict,
     xlab = "Observed",
     ylab = "Predicted",
     xlim = c(0, 12),
     ylim = c(0, 12),
     pch = 16)
abline(0, 1)
```

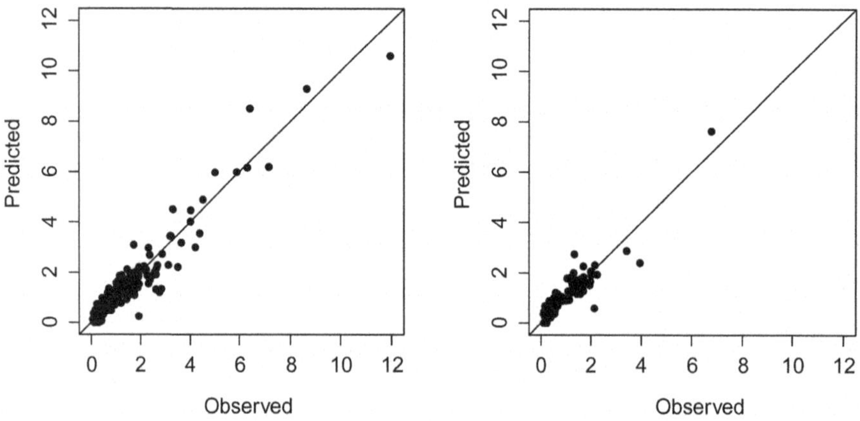

Fig. 9.16 Scatterplot of observed against predicted values of the total carbon. Predictions are made using a cubist model with only two rules

Derive some accuracy measures, for the example using the `eval` function and specifying the argument `obj = "quant"` for continuous variables:

```
# accuracy measures for calibration
soilspec::eval(datC$TotalCarbon, soilCCubist2Predict, obj = "quant")
```

```
##      ME RMSE   r2   R2 rhoC  RPD RPIQ
## 1 -0.01  0.4 0.86 0.93 0.96 3.76 2.91
```

and for validation.

```
# accuracy measures for validation
soilspec::eval(datV$TotalCarbon, soilVCubist2Predict, obj = "quant")
```

```
##     ME RMSE   r2   R2 rhoC  RPD RPIQ
## 1 0.06 0.39 0.79 0.82 0.91 2.38 3.08
```

We can also infer which wavelengths are useful in the model based on the model summary. The following plot shows which variables are important as predictors and as 'conditions' (Fig. 9.17).

```
# plot the variables used as predictors
plot(soilCCubistModel$usage[, 3], soilCCubistModel$usage[, 2],
     type = "h",
     col = "plum",
     xlab = "Wavelength /nm",
     ylab = "Model usage /%")

# plot the conditions in blue
lines(soilCCubistModel$usage[, 3], soilCCubistModel$usage[, 1],
      type = "h",
      col = "blue")  # see which variables are important

# add to the existing plot
par(new = T)

# add a spectra for visualization
plot(colnames(datC$spcAMovav), datC$spcAMovav[1, ],
     axes = F,
     ylim = c(-2, 2),
     xlab = " ", ylab = " ",
     type = "l",
     main = " ",
     xlim = c(500, 2450))
```

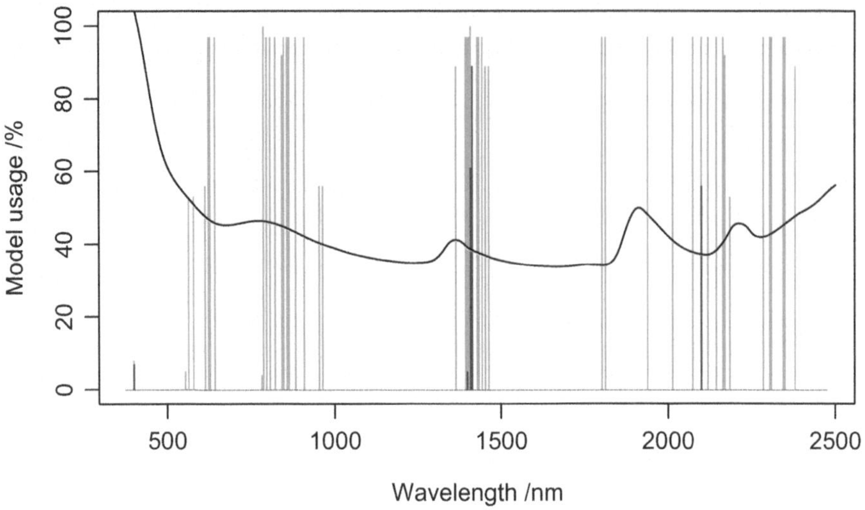

Fig. 9.17 Important variables of the cubist model to predict total carbon. The black line is an example pre-processed absorbance spectrum from the `datsoilspc` dataset, the pink vertical lines are the variables (wavelength) used as predictors in the cubist model, and the vertical blue lines are the cubist model conditions

9.2.4 Random Forest

Random forest (RF) is a widely used algorithm for data science applications in a broad array of scientific domains. A RF model is an ensemble of decision trees organized as set of structured classifiers or trees and can be used for both regression and classification purposes (Breiman 2001). The advantage of RF over its much simpler data-mining counterpart, the CART (Classification and Regression Tree) model, is its ensemble approach. Rather than a single tree in the case of CART, the RF model has many trees, where each is constructed using different perturbations of the data – in terms of both calibration cases and explanatory variables. During RF model development of the ensemble trees, two-thirds of cases are sampled (bootstrap sampling with replacement) and are used to grow a regression tree, and the other one-third is used to perform a cross-validation in parallel with the calibration step. These samples are called out-of-bag samples (oob samples) which are used to obtain an estimate of the model performance. For regression the final prediction is the average of the individual tree or classifier outputs, whereas in classification the trees vote by majority on the correct classification (mode). You might note the similarity of this cross-validation procedure in the earlier described section about the bootstrap PLSR model.

Random forest has three tuning parameters: mtry, number of trees and minimum node size. The first parameter mtry is the number of input variables that are randomly selected for each bootstrap, which can range from 1 to n (sample size).

The second parameter is the number of trees, which must be sufficiently large for the oob error stabilization. In general, 500 trees are sufficient, but if a large number of trees are chosen, the results will not differ significantly, but more time will be necessary to model fitting. The last tuning parameter is the minimum node size, which determines the minimum size of nodes which no split will be attempted; the default value is 5 for regression and 1 for classification.

The RF model is a natural candidate for soil spectral inference work because of the high dimension status of the soil spectral wavelengths, for example, with vis-NIR or MIR data. Ideally it would be used in situations where the number of spectra is also large, as this model has a tendency to overfit, particularly if the tunable parameters previously mentioned are not optimized. The issue here is about generalization when the model is extended to new data, which is not a strong feature of the RF model. Naturally its wide use in data science has also resulted in some researchers experimenting with this model approach for soil spectral data with good results, e.g. Santana et al. (2018) and Hobley et al. (2017) as some relatively recent examples.

In this example, we use the `randomForest` package.

```
# load the required package
require(randomForest)

# prepare the data, the column name cannot be numeric, add 'spec.' in front
datCSub <- data.frame(TotalCarbon = datC$TotalCarbon, datC$spcAMovav)
colnames(datCSub) <- c("TotalCarbon", paste0("spec.", colnames(datC$spcAMovav)))


# run random forest algorithm
soilCRFModel <- randomForest(TotalCarbon ~ .,
                             data = datCSub,
                             ntree = 1000,
                             mtry = 10,
                             importance = TRUE,
                             na.action = na.omit)


# summary of the model
soilCRFModel
```

```
##
## Call:
##  randomForest(formula = TotalCarbon ~ ., data = datCSub, ntree = 1000,
                                  mtry = 10, importance = TRUE, na.action = na.omit)
##                Type of random forest: regression
##                      Number of trees: 1000
## No. of variables tried at each split: 10
##
##           Mean of squared residuals: 0.4582793
##                     % Var explained: 79.18
```

The `randomForest` function saves the important variables that were used to explain the variance of the soil property. For this, we specified `importance = TRUE`, and we can now plot the important variables of the `soilCRFModel` model. Two methods are proposed, the change in terms of MSE or in terms of node purity. The reader is redirected to statistical learning book (e.g. Friedman et al. 2001) for more information on these methods. Figure 9.18 shows the most important variables sorted in order of importance.

```
# plot the most important bands of the spectra
varImpPlot(soilCRFModel,
           main = " ")
```

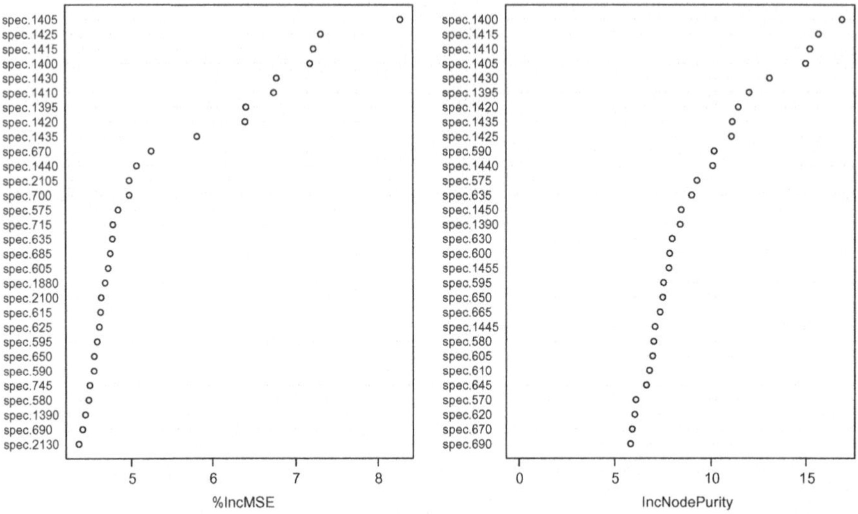

Fig. 9.18 Important variables used by the random forest model to explain the variability of the total carbon. The two methods are the mean decrease in accuracy (left) and mean decrease in node impurity (right). The first variables (bands) are the most important ones

Using the fitted RF model, we can generate predictions on the validation set and validate the predictions (Fig. 9.19).

```
# predict on the calibration data
soilCRFPred <- predict(soilCRFModel, datCSub)

# prepare the validation data
datVSub <- data.frame(TotalCarbon = datV$TotalCarbon, datV$spcAMovav)
colnames(datVSub) <- c("TotalCarbon", paste0("spec.", colnames(datV$spcAMovav)))

# predict on the validation data
soilVRFPred <- predict(soilCRFModel, datVSub)

par(mfrow = c(1, 2))

# plot calibration
plot(datC$TotalCarbon, soilCRFPred,
     xlab = "Observed",
     ylab = "Predicted",
     xlim = c(0, 12),
     ylim = c(0, 12),
     pch = 16)
abline(0, 1)

# plot validation
plot(datV$TotalCarbon, soilVRFPred,
```

```
     xlab = "Observed",
     ylab = "Predicted",
     xlim = c(0, 12),
     ylim = c(0, 12),
     pch = 16)
abline(0, 1)
```

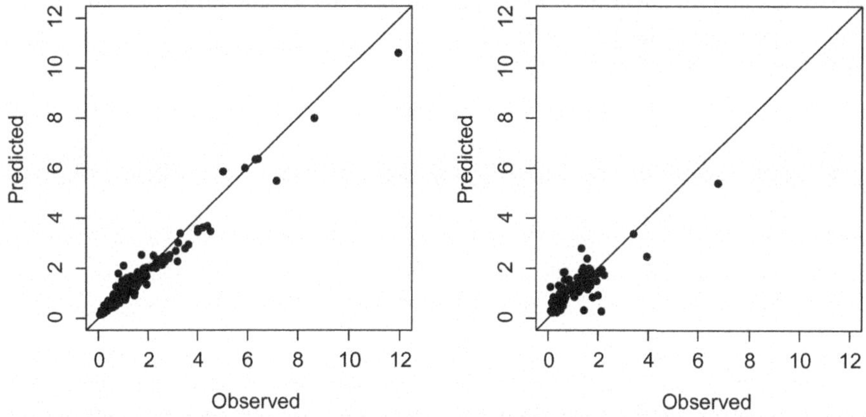

Fig. 9.19 Scatterplot of observed against predicted values of the total carbon. Predictions are made using a random forest model

Derive some accuracy measures, for the example using the `eval` function and specifying the argument `obj = "quant"` for continuous variables:

```
# accuracy measures for calibration
soilspec::eval(datC$TotalCarbon, soilCRFPred, obj = "quant")
```

```
##    ME RMSE   r2   R2 rhoC  RPD RPIQ
## 1  0   0.3 0.94 0.96 0.98 5.03 3.89
```

and for validation.

```
# accuracy measures for validation
soilspec::eval(datV$TotalCarbon, soilVRFPred, obj = "quant")
```

```
##     ME RMSE   r2   R2 rhoC RPD RPIQ
## 1 0.1 0.52 0.57 0.69 0.81 1.8 2.33
```

9.2.5 Memory-Based Learning

Memory-based learning (MBL) is a local calibration algorithm presented in Ramirez-Lopez et al. (2013a) and implemented in the package `resemble`. MBL

has been developed to deal with complex, often continental or global, soil (infrared) spectral datasets. Instead of building a single (global) function to link the soil property and the spectra (as in PLSR or RF), the spectra are split into a number of subsets sharing similar spectral characteristics. In this sense, this technique is closely related to Chap. 7 because it makes use of distance metrics in the principal component space to select the closest points for building a local model. Subsequently, MBL can deal with complex non-linear relationships between the spectra and a particular soil property.
MBL works in the following stages:

1. Build a p-dimensional space of the spectra where p is the number of principal components.
2. For each point of the validation dataset, select its k-nearest neighbours from the calibration sample.
3. Fit a local model for each point of the validation dataset using its nearest neighbours spectra found in the calibration dataset. Several models are available for the fit.

Some aspects are therefore particularly important in any MBL algorithm. One must choose the similarity/dissimilarity metric used to select the closest point in the spectra space (see also Chap. 7). A second important point is the number of neighbours that one wants to use in the fitting process. This number must be sufficiently large to build a realistic model, but increasing too much the number of points might also decrease prediction accuracy. The package `resemble` provides options to optimize the number of neighbours.

In the function `mbl`, the user must then decide:

- the similarity metric `diss_method`, in this example the Mahalanobis distance in the PC space (`diss_method = "pca"`).
- the choice of the optimal number of PCs under the argument `pc_selection`; in this example, the PCs are selected based on the cumulative amount of variance explained.
- the model for fitting, in our example the weighted average PLS model (`method = "local_fit_wapls(min_pls_c = 4, max_pls_c = 17)"`).
- the validation method used, in this example the leave-nearest-neighbour-out cross-validation (`validation_type = "NNv"` in the `mbl_control` argument).

We further decide a sequence of nearest neighbours `k` to test (in order to find the optimal number of neighbours in local model fitting). Here we test a sequence from 20 up to 120 in steps of 10.

```
# define the sequence of neighbours
k2t <- seq(from = 20, to = 120, by = 10)
```

We can now perform the model fitting using MBL and the `mbl` function. We use the object `control` which contains the parameters of the `mbl` function. We make use of a regression function called weighted average partial least square (`wapls1`). It uses multiple models generated by multiple PLS components (i.e. between a minimum and a maximum number of PLS components). At each local partition, the final predicted value is a weighted average of all the predicted values generated by the multiple PLS models. See the package documentation for more details.

```
# load the required package
require(resemble)

# maximum cumulative variance explained to be retained by the PCs
maxexplvar <- 0.99

# run the mbl algorithm
mblResults1 <- mbl(Xr = datC$spcAMovav,
                   Yr = datC$TotalCarbon,
                   # we assume we do not know the total carbon content of the testing
                     dataset
                   Yu = NULL,
                   Xu = datV$spcAMovav,
                   diss_method = "pca",
                   control = mbl_control(validation_type = "NNv"),
                   diss_usage = "none",
                   k = k2t,
                   # define the number of minimum and maximum components for "wapls1"
                   method = local_fit_wapls(min_pls_c = 4, max_pls_c = 17),
                   pc_selection = list("cumvar", maxexplvar),
                   scale = FALSE, center = TRUE)
```

We can summarize the information derived and plot the results.

```
# print a summary of the model
mblResults1
```

```
##
## Call:
##
## mbl(Xr = datC$spcAMovav, Yr = datC$TotalCarbon, Xu = datV$spcAMovav,
##     Yu = NULL, k = k2t, method = local_fit_wapls(min_pls_c = 4,
##         max_pls_c = 17), diss_method = "pca", diss_usage = "none",
##     pc_selection = list("cumvar", maxexplvar), control = mbl_control(validation_
                                                                    type = "NNv"),
##     center = TRUE, scale = FALSE)
##
## _______________________________________________________
##
##  Total number of observations predicted: 98
## _______________________________________________________
##
##  Nearest neighbor validation statistics
##
##       k  rmse st_rmse    r2
##  1:  20 0.530  0.0833 0.784
##  2:  30 0.444  0.0699 0.843
##  3:  40 0.523  0.0823 0.846
##  4:  50 0.542  0.0853 0.850
##  5:  60 0.527  0.0829 0.848
##  6:  70 0.556  0.0874 0.829
##  7:  80 0.549  0.0863 0.828
##  8:  90 0.568  0.0894 0.834
##  9: 100 0.549  0.0863 0.850
```

```
## 10: 110 0.550  0.0864 0.848
## 11: 120 0.535  0.0841 0.851
## ________________________________________________________
```

```
# plot the RMSE against number of neighbours
matplot(mblResults1$validation_results$nearest_neighbor_validation$k,
        mblResults1$validation_results$nearest_neighbor_validation$rmse,
        type = "b",
        xlab = "K-neighbours",
        ylab = "RMSE",
        pch = 1,
        col = "dodgerblue")
```

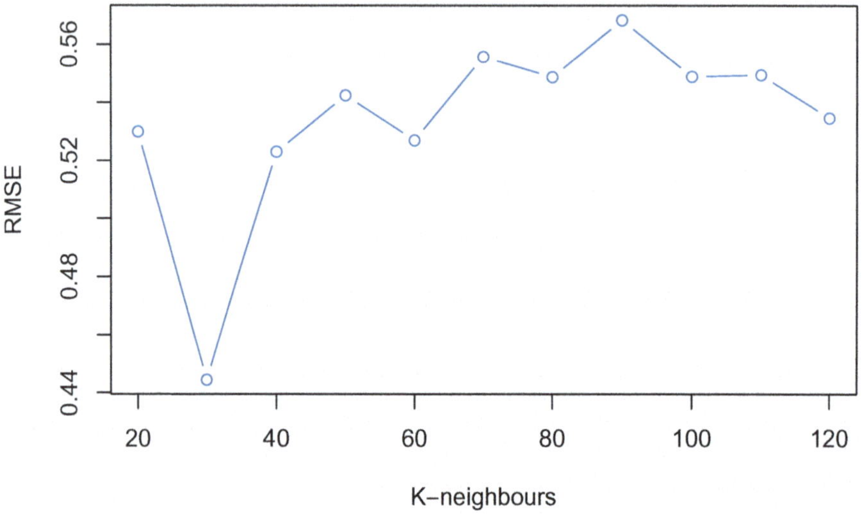

Fig. 9.20 Values of the root mean square error against number of neighbours in the memory-based learning algorithm

Figure 9.20 shows the number of neighbours against the RMSE. The optimal number of neighbours that minimizes the RMSE is 30.

```
# rmse values
rmseMBL <- mblResults1$validation_results$nearest_neighbor_validation$rmse

# minimum rmse value
minRmseMBL <- min(mblResults1$validation_results$nearest_neighbor_validation$rmse)

# number of neighbours values
neighNumber <- mblResults1$validation_results$nearest_neighbor_validation

# select the optimal number of neighbours
optNn <- neighNumber[rmseMBL == minRmseMBL,]$k
```

Instead of assuming that the total carbon content in the validation set is unknown (the `NULL` in the `Yu` argument), we can use the actual total carbon content values of this dataset. Note that they are not used at all during computations and they are only used for validation purposes (i.e. comparing what it was predicted to the actual values).

```
# run the mbl algorithm specifying the Yu argument
mblResults1Val <- mbl(Xr = datC$spcAMovav,
                      Yr = datC$TotalCarbon,
                      Yu = datV$TotalCarbon,
                      Xu = datV$spcAMovav,
                      diss_method = "pca",
                      control = mbl_control(validation_type = "NNv"),
                      diss_usage = "none",
                      k = optNn,
                      # define the number of minimum and maximum components for "wapls1"
                      method = local_fit_wapls(min_pls_c = 4, max_pls_c = 17),
                      pc_selection = list("cumvar", maxexplvar),
                      scale = FALSE, center = TRUE)
```

We can plot the predicted and observed values of the total carbon as done before for the other calibration algorithms (Fig. 9.21).

```
# plot validation
plot(datV$TotalCarbon, mblResults1Val$results$k_30$pred,
     xlab = "Observed",
     ylab = "Predicted",
     xlim = c(0, 12),
     ylim = c(0, 12),
     pch = 16)
abline(0, 1)
```

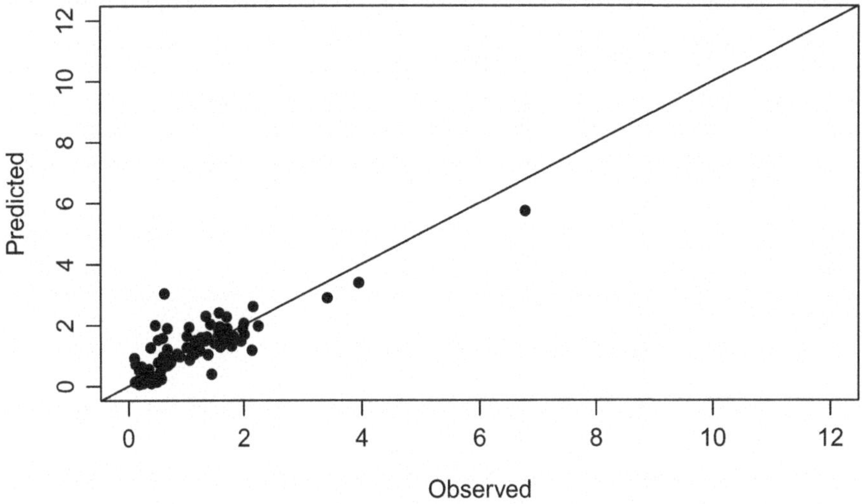

Fig. 9.21 Scatterplot of observed against predicted values of the total carbon. The predictions are made by MBL using a weighted local PLS model

Derive some accuracy measures, for the example using the `eval` function and specifying the argument `obj = "quant"` for continuous variables:

```
# accuracy measures for validation
soilspec::eval(datV$TotalCarbon, mblResults1Val$results$k_30$pred, obj = "quant")
```

```
##       ME RMSE   r2   R2 rhoC  RPD RPIQ
## 1 0.13  0.5 0.65 0.71 0.85 1.86 2.42
```

9.3 Models for Categorical Variables

Soil properties such as organic carbon or clay are continuous. Some other soil properties or attributes are categorical, that is, they are assigned to a finite number of groups. Discrete values can also be considered as categorical if their number is limited. Examples of categorical soil properties are soil texture classes, mineral categories or soil types.

For prediction of categorical soil properties, the models need to be adapted. Since predicting categorical properties with spectroscopy is relatively rare in soil science, we will present two models, which are adaptations of models already presented in the previous section on modelling continuous properties.

As an example, we use the continuous values of clay, silt and sand provided in the book-associated Geeves dataset and convert it to soil texture classes. The dataset originates from Australia (see also Chap. 3), hence the use of the Australian soil texture triangle (Northcote 1971; National Committee on Soil and Terrain (Australia) and CSIRO Publishing 2009) to define the class names. The Australian soil texture triangle has 11 classes. We show below how to convert the soil clay, silt and sand content to soil texture classes using the `soiltexture` R package.

We start by plotting the Australian soil texture triangle and the points from our dataset (Fig. 9.22).

```
# load the require package
require(soiltexture)

# we create the tables
datCsub <- data.frame(CLAY = datC$clay,
                      SILT = datC$silt,
                      SAND = datC$sand)
datVsub <- data.frame(CLAY = datV$clay,
                      SILT = datV$silt,
                      SAND = datV$sand)

# we normalize the data so that the sum of clay, silt and sand is 100
datCsub <- TT.normalise.sum(tri.data = datCsub)
datVsub <- TT.normalise.sum(tri.data = datVsub)

# plot the soil texture triangle for the calibration data
TextPlot <- TT.plot(class.sys = "AU.TT",
                    tri.data = datCsub,
                    main = " ",
                    pch = 16,
```

```
                    cex.axis = 0.7,
                    cex.lab = 0.7)

# add to the soil texture triangle the validation data
TT.points(tri.data = datVsub,
          geo = TextPlot,
          col = "red",
          pch = 16)
```

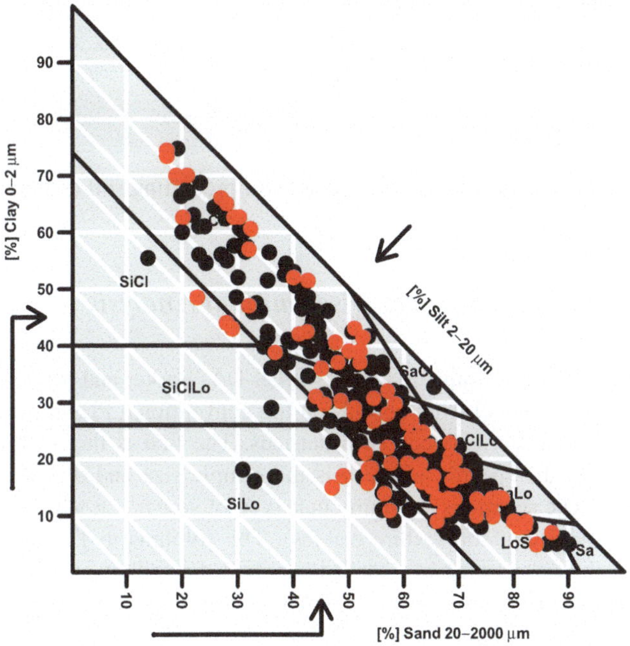

Fig. 9.22 Australian soil texture triangle with location of the calibration (black dots) and validation (red dots) soil texture classes within the triangle

We can assign the name of the Australian soil texture class to the calibration and validation datasets. By default, the points that are at the border of two classes are assigned the name of the two classes. For the demonstration, we remove these points. In most cases, however, it would be more judicious to make further diagnostics on the soil samples and to assign a class to these samples.

```
# make soil texture classes for the calibration dataset
datC$textclass <- TT.points.in.classes(tri.data = datCsub,
                                       class.sys = "AU.TT",
                                       text.tol = 1,
                                       PiC.type = "t",
                                       collapse    = "_")
```

```
# The points that fall into two classes are removed.
datC <- datC[-grep("_", datC$textclass),]

# make soil texture classes for the validation dataset
datV$textclass <- TT.points.in.classes(tri.data = datVsub,
                                       class.sys = "AU.TT",
                                       text.tol = 1,
                                       PiC.type = "t",
                                       collapse    = "_")

# The points that fall into two classes are removed.
datV <- datV[-grep("_", datV$textclass),]

# show the derived categories of soil classes in the calibration dataset
unique(datC$textclass)
```

```
## [1] "Cl"    "Lo"    "SiLo"   "SaLo"   "ClLo"   "LoSa"   "SiClLo" "SaClLo"
## [9] "SiCl"  "SaCl"
```

Now we will show how to use two models for predicting soil texture classes.

9.3.1 Partial Least Squares Discriminant Analysis

Partial least squares discriminant analysis (PLS-DA, Barker and Rayens 2003) is a linear classification model which builds on the conventional PLS regression. PLS-DA is a PLS regression between two matrices. The first is the matrix **X** of size $n \times b$ containing the spectra, where n is the sample size and b is the number of spectral bands. The second is a dummy matrix **Y** of size $n \times c$ where c is the total number of classes (e.g. soil texture classes). The columns in Y represent a class membership, that is, a value of 1 or 0 is applied depending on whether the soil sample belongs to the class. The PLS-DA model is then calibrated like PLS, by a linear regression between the rotated scores of the **X** and **Y** matrices principal components. The calibrated model predicts a value between 0 and 1, which is assigned to the closest class by a softmax function. A discussion on PLS-DA is further provided by Brereton and Lloyd (2014).

In this example, we use the `caret` package which provides functionalities for cross-validation.

```
# load the require package
require(caret)

# create a subset of the calibration dataset
datCSub <- data.frame(soiltext = datC$textclass, datC$spcAMovav)
colnames(datCSub) <- c("textclass", paste0("spec.", colnames(datC$spcAMovav)))

# we do a k-fold cross validation repeated three times
ctrl <- trainControl(method = "repeatedcv",
                     repeats = 3)

# set the seed
set.seed(123)

# build the PLS-DA model using the caret package
soilCPlsdaModel <- train(textclass ~ .,
```

```
                          data = datCSub,
                          method = "pls",
                          preProc = c("center", "scale"),
                          metric = c("Accuracy"),
                          tuneLength = 30,
                          trControl = ctrl)
```

We can display a summary of the fitted PLS-DA model.

```
# summary of the fitted model
soilCPlsdaModel
```

```
## Partial Least Squares
##
## 290 samples
## 421 predictors
##  10 classes: 'Cl', 'ClLo', 'Lo', 'LoSa', 'SaCl', 'SaClLo', 'SaLo', 'SiCl', 'SiClLo',
                              'SiLo'
##
## Pre-processing: centered (421), scaled (421)
## Resampling: Cross-Validated (10 fold, repeated 3 times)
## Summary of sample sizes: 260, 258, 260, 261, 263, 263, ...
## Resampling results across tuning parameters:
##
##   ncomp  Accuracy   Kappa
##    1     0.5534921  0.3461652
##    2     0.5593630  0.3558506
##    3     0.5593189  0.3567091
##    4     0.5650486  0.3656058
##    5     0.5650045  0.3654928
##    6     0.5650045  0.3660005
##    7     0.5708395  0.3736145
##    8     0.5744451  0.3835748
##    9     0.5823474  0.3970096
##   10     0.5640763  0.3746830
##   11     0.5738831  0.3908864
##   12     0.5758560  0.3937251
##   13     0.5850061  0.4117775
##   14     0.5668808  0.3889140
##   15     0.5726905  0.3977221
##   16     0.5853583  0.4178228
##   17     0.5868221  0.4213265
##   18     0.5960500  0.4360405
##   19     0.6052095  0.4509891
##   20     0.6046920  0.4543307
##   21     0.6234867  0.4825286
##   22     0.6270725  0.4894305
##   23     0.6304389  0.4948148
##   24     0.6418556  0.5108486
##   25     0.6484922  0.5213268
##   26     0.6416819  0.5124769
##   27     0.6437369  0.5150439
##   28     0.6414408  0.5116607
##   29     0.6357714  0.5035562
##   30     0.6295964  0.4961511
##
## Accuracy was used to select the optimal model using the largest value.
## The final value used for the model was ncomp = 25.
```

The `train` function of the `caret` package automatically selects the optimal number of principal components based on the overall accuracy evaluated by the cross-validation subset left out (Fig. 9.23).

```r
# plot the cross-validation results
plot(soilCPlsdaModel)
```

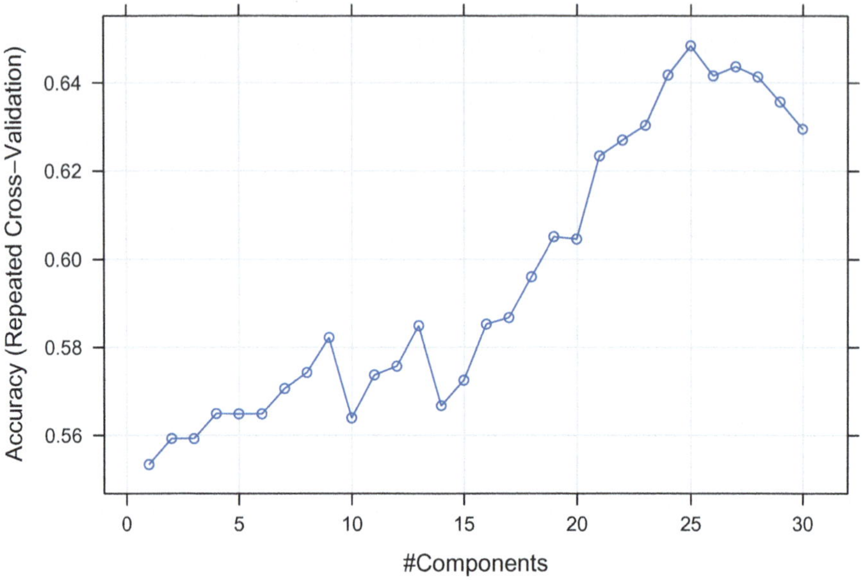

Fig. 9.23 Number of components used in the partial least squares discriminant analysis model against accuracy computed from a repeated cross-validation

The `train` function selected 25 components as the optimal number. The fitted PLS-DA model can now be applied to unseen data.

```r
# create a validation dataset
datVSub <- data.frame(soiltext = datV$textclass, datV$spcAMovav)
colnames(datVSub) <- c("textclass", paste0("spec.", colnames(datV$spcAMovav)))

# show the probability of each class
head(predict(soilCPlsdaModel,
             newdata = datVSub,
             type="prob"))
```

```
##          Cl       ClLo         Lo       LoSa       SaCl     SaClLo       SaLo
## 1 0.1201323 0.16463553 0.07922130 0.09377720 0.08920005 0.08531693 0.08447906
## 2 0.2054752 0.10797734 0.06762277 0.10687099 0.08748491 0.08216171 0.07374156
## 3 0.1930218 0.09038021 0.08228958 0.09765407 0.08766723 0.08802700 0.09041075
## 4 0.1163598 0.07402499 0.11089104 0.14420159 0.08809790 0.08977505 0.10009583
## 5 0.2927797 0.05545218 0.09967716 0.07337192 0.08037484 0.08205305 0.07691914
## 6 0.1025771 0.06678866 0.23378266 0.08995787 0.08500366 0.08440278 0.08189733
##         SiCl     SiClLo       SiLo
## 1 0.09069487 0.09012774 0.10241505
## 2 0.08595896 0.08518364 0.09752289
## 3 0.08966126 0.09750844 0.08337970
## 4 0.08502138 0.08641169 0.10512070
## 5 0.08054953 0.07722688 0.08159563
## 6 0.08272078 0.08220722 0.09066194
```

Here we can see that the output is a number between 0 and 1. As mentioned previously, the PLS-DA model returns a probability-like value of the assignment to a class. For example, the fifth soil sample is more likely to belong to the class `Cl` (probability of 0.20) than it is to the other classes (probability below 0.1 in all other classes). To obtain the final results, the softmax function is applied on the probability values to return the single value, that is, the membership to a class. This is done by default in the `caret` package.

We can now visualize the prediction and validate against measured value of the classes.

```
# predict the classes on the validation dataset using the fitted PLS-DA model
plsClasses <- predict(soilCPlsdaModel, newdata = datVSub)

# create a confusion matrix of the predicted versus observed soil texture classes
confMat <- confusionMatrix(datVSub$textclass, plsClasses)

# show confusion matrix computed on the validation dataset
confMat$table
```

```
##             Reference
## Prediction Cl ClLo Lo LoSa SaCl SaClLo SaLo SiCl SiClLo SiLo
##     Cl     21    1  2    0    0      0    0    0      0    0
##     ClLo    6    5  7    0    0      0    0    0      0    0
##     Lo      0    2 24    0    0      0    0    0      0    0
##     LoSa    0    0  5    7    0      0    0    0      0    0
##     SaCl    0    0  0    0    0      0    0    0      0    0
##     SaClLo  0    1  0    0    0      0    0    0      0    0
##     SaLo    0    1  3    2    0      0    0    0      0    0
##     SiCl    2    0  1    0    0      0    0    0      0    0
##     SiClLo  0    0  0    0    0      0    0    0      0    0
##     SiLo    0    0  6    1    0      0    0    0      0    0
```

```
# show accuracy and Cohen's kappa
confMat$overall[1:2]
```

```
##  Accuracy     Kappa
## 0.5876289 0.4584787
```

The confusion matrix `confMat` shows that there is on average a good agreement between predicted and measured soil texture classes. For example, 22 of the reference `Cl` values are correctly classified, and 11 are assigned to a different class, 8 of which are to the most similar class (`ClLo`). The overall accuracy and Cohen's kappa values show that the predictions are accurate.

9.3.2 Random Forest

Random forest applied to categorical variables is similar to the same model applied to continuous variables. Note that parameter tuning is not performed in this example and that, for example, mtry is held constant. The reader will find a large number of examples on how to make parameter tuning in the package vignette. Parameter tuning will be a key element of the modelling, in particular to avoid overfitting of the model or to avoid an excessive number of trees without compromising on prediction accuracy.

```
# we start by defining the control parameters for the caret function
# we do a k-fold cross validation repeated three times
trainControl <- trainControl(method = "repeatedcv",
                             number = 10,
                             repeats = 3)

# define the random forest parameter mtry (to its default)
mtry <- sqrt(ncol(datCSub))

# hold the mtry parameter constant (not parameter tuning)
tunegrid <- expand.grid(.mtry = mtry)

# set the seed
set.seed(123)

# build the random forest model using the caret package
soilCRFModel <- train(textclass ~ .,
                      data = datCSub,
                      method = "rf",
                      metric = c("Accuracy"),
                      tuneGrid = tunegrid,
                      trControl = trainControl,
                      verbose = FALSE)

# print a summary of the model
print(soilCRFModel)
```

```
## Random Forest
##
## 290 samples
## 421 predictors
##  10 classes: 'Cl', 'ClLo', 'Lo', 'LoSa', 'SaCl', 'SaClLo', 'SaLo', 'SiCl', 'SiClLo',
                  'SiLo'
##
## No pre-processing
## Resampling: Cross-Validated (10 fold, repeated 3 times)
## Summary of sample sizes: 260, 258, 260, 261, 263, 263, ...
## Resampling results:
##
##   Accuracy   Kappa
##   0.6269545  0.5055345
##
## Tuning parameter 'mtry' was held constant at a value of 20.54264
```

Let us now apply the calibrated random forest model to the validation data and display the overall accuracy and Cohen's kappa statistics. Note that the same statistics can be obtained by using the book-associated package `soilspec` with the `eval` function.

```
# predict the classes on the validation dataset using the fitted random forest model
RFClasses <- predict(soilCRFModel,
                     newdata = datVSub)

# create a confusion matrix of the predicted versus observed soil texture classes
confMat <- confusionMatrix(datVSub$textclass, RFClasses)

# show confusion matrix computed on the validation dataset
confMat$table
```

```
##          Reference
## Prediction Cl ClLo Lo LoSa SaCl SaClLo SaLo SiCl SiClLo SiLo
##     Cl     22    2  0    0    0      0    0    0      0    0
##     ClLo    8    8  2    0    0      0    0    0      0    0
##     Lo      0    1 20    4    0      0    1    0      0    0
##     LoSa    0    0  0   12    0      0    0    0      0    0
##     SaCl    0    0  0    0    0      0    0    0      0    0
##     SaClLo  0    1  0    0    0      0    0    0      0    0
##     SaLo    0    0  6    0    0      0    0    0      0    0
##     SiCl    3    0  0    0    0      0    0    0      0    0
##     SiClLo  0    0  0    0    0      0    0    0      0    0
##     SiLo    0    0  4    1    0      0    0    0      0    2
```

```
# show accuracy and Cohen's kappa
confMat$overall[1:2]
```

```
##  Accuracy     Kappa
## 0.6597938 0.5641933
```

9.4 Soil Spectral Inference Systems

Another way to estimate soil properties is to combine the spectra with pedotransfer functions (PTF, Bouma 1989), in a soil spectral inference system (McBratney et al. 2006). Several soil physical, chemical and biological properties cannot be estimated directly from the spectra, as it done in Sects. 9.2 and 9.3. The reasons are that i) some soil properties do not have a clear spectral response in the spectrum and that ii) the development of calibration functions of a soil property from soil spectral libraries is not always possible due, for example, to budget constraints. McBratney et al. (2006) thus proposed a two-step approach to estimate a large range of soil properties by combining spectroscopy and PTFs, as follows:

- Step 1, calibration: a multivariate model is built between the spectra and the measured values of some basic soil properties (that have been shown to demonstrate a spectral response in the spectral region of interest). In the infrared, the basic soil properties are, for example, clay, silt, sand, organic carbon, pH and cation exchange capacity. This step can be implemented using one of the methods presented in Sects. 9.2 or 9.3. When a model is calibrated, it can be used to predict soil properties of a soil sample where only the spectrum is available.
- Step 2, inference: the basic soil properties estimated in Step 1 are used as input in a PTF to predict a set of different soil properties, such as the permanent soil wilting point or field capacity. The PTF can be found in the literature or derived using a large soil database. Ideally, a PTF developed on similar soil is used to derive the soil properties. A number of PTFs have been published, for example, by McBratney et al. (2002) or Pachepsky and Rawls (2004). The PTFs can be concatenated into a network structure to create an inference system that estimates the target soil property or properties and also propagates the uncertainty of the estimate.

So one of the main features of soil spectral inference systems is the quantification and propagation of uncertainty. Model uncertainty, for example, can be quantified using an ensemble model by bootstrap and aggregating, for which an example using partial least squares regression is provided in Sect. 9.2.2.

Since Step 1 is already presented in Sects. 9.2 or 9.3, we do not repeat it here and make a simple example of Step 2. Note also that uncertainty quantification and propagation is discussed elsewhere in the literature (e.g. in Tranter et al. 2010, Van der Klooster et al. 2011 or Brodský et al. 2013).

Take the following PTF from Rab et al. (2011) to estimate the volumetric field capacity (FC, m^3/m^3, %) as a function of basic soil properties. We chose this PTF because it has been derived using data from a case study in Australia, for an area in which the soils are similar to those of the Geeves dataset. The Geeves dataset is provided by the book-associated `soilspec` package. The PTF makes use of the soil clay and silt content.

$$Field\ capacity = 7.759 + 0.7165 \times clay + 0.9708 \times silt - (0.01729 \times clay) \times silt \tag{9.16}$$

We can use the Geeves dataset provided in the `soilspec` package. It contains values of the soil clay and silt content.

```
# load the required package
require(soilspec)

# load the data
data("datsoilspc")

# show the available soil properties
colnames(datsoilspc)
```

```
## [1] "clay"          "silt"          "sand"          "TotalCarbon" "spc"
```

Note here the soil properties are available in our dataset but that in many cases they can be estimated using the spectra (Step 1) or using a soil spectral library (Viscarra-Rossel et al. 2008). We can now derive the field capacity using the basic soil properties clay and silt (Fig. 9.24).

```
# estimate field capacity using a PTF
FC <-  7.759 +
  0.7165*datsoilspc["clay"] + 0.9708*datsoilspc["silt"]  -
  (0.01729*datsoilspc["clay"] )*datsoilspc["silt"]

# plot the distribution of the estimated field capacity
boxplot(FC,
        ylab = "Field capacity")
```

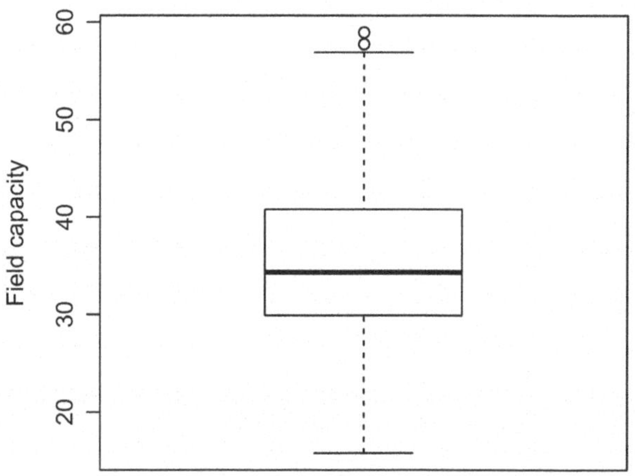

Fig. 9.24 Boxplot of the estimated values of the field capacity (in m^3/m^3, %) of the Geeves dataset

References

Barker M, Rayens W (2003) Partial least squares for discrimination. J Chemometr J Chemometrics Soc 17:166–173

Batten GD (1998) Plant analysis using near infrared reflectance spectroscopy: the potential and the limitations. Aust J Exp Agric 38:697–706

Bellon-Maurel V, Fernandez-Ahumada E, Palagos B, Roger J-M, McBratney AB (2010) Critical review of chemometric indicators commonly used for assessing the quality of the prediction of soil attributes by NIR spectroscopy. TrAC Trends Anal Chem 29:1073–1081

Bouma J (1989) Using soil survey data for quantitative land evaluation. In: Advances in soil science. Springer, Berlin, pp 177–213

Breiman L (2001) Random forests. Mach Learn 45:5–32

Breiman L (1996) Bagging predictors. Mach Learn 24:123–140

Brereton RG, Lloyd GR (2014) Partial least squares discriminant analysis: taking the magic away. J Chemometr 28:213–225

Brodský L, Vasat R, Klement A, Zadorova T, Jaksik O (2013) Uncertainty propagation in VNIR reflectance spectroscopy soil organic carbon mapping. Geoderma 199:54–63

Brus DJ, Kempen B, Heuvelink GBM (2011) Sampling for validation of digital soil maps. Eur J Soil Sci 62:394–407

Chang C-W, Laird DA, Mausbach MJ, Hurburgh CR (2001) Near-infrared reflectance spectroscopy–principal components regression analyses of soil properties. Soil Sci Soc Am J 65:480–490

Cohen J (1960) A coefficient of agreement for nominal scales. Educ Psychol Meas 20:37–46

Esbensen KH, Geladi P, Larsen A (2014) The RPD myth.... NIR News 25:24–28

Friedman J, Hastie T, Tibshirani R (2001) The elements of statistical learning. Springer series in statistics, New York

Geeves GW, Cresswell HP, Murphy BW, Gessler PI, Chartres CJ, Little IP, Bowman GM (1994) Physical, chemical and morphological properties of soils in the wheat-belt of southern NSW and northern Victoria. NSW Department of Conservation; Land Management/CSIRO Division of Soils Occasional Report, CSIRO

Hobley EU, Brereton AJLEG, Wilson B (2017) Soil charcoal prediction using attenuated total reflectance mid-infrared spectroscopy. Soil Res 55:86–92

Janssen PHM, Heuberger PSC (1995) Calibration of process-oriented models. Ecol Model 83:55–66
Kuhn M, Weston S, Keefer C, Coulter N (2012) Cubist models for regression. R package Vignette R package version 00 18
Lawrence I, Lin K (1989) A concordance correlation coefficient to evaluate reproducibility. Biometrics 45:255–268
McBratney AB, Minasny B, Cattle SR, Vervoort RW (2002) From pedotransfer functions to soil inference systems. Geoderma 109:41–73
McBratney AB, Minasny B, Viscarra-Rossel RA (2006) Spectral soil analysis and inference systems: A powerful combination for solving the soil data crisis. Geoderma 136:272–278
Mevik B-H, Segtnan VH, Næs T (2004) Ensemble methods and partial least squares regression. J Chemometr J Chemometr Soc 18:498–507
Minasny B, McBratney AB (2013) Why you don't need to use RPD. Pedometron 33:14–15
Minasny B, McBratney AB (2008) Regression rules as a tool for predicting soil properties from infrared reflectance spectroscopy. Chemom Intell Lab Syst 94:72–79
Nash JE, Sutcliffe JV (1970) River flow forecasting through conceptual models part I—A discussion of principles. J Hydrol 10:282–290
National Committee on Soil and Terrain (Australia) and CSIRO Publishing (2009) Australian soil and land survey field handbook. CSIRO Publishing
Ng W, Minasny B, Malone BP, Sarathjith MC, Das BS (2019) Optimizing wavelength selection by using informative vectors for parsimonious infrared spectra modelling. Comput Electron Agric 158:201–210
Northcote KH (1971) Factual key for the recognition of Australian soils
Nussbaum M, Walthert L, Fraefel M, Greiner L, Papritz A (2017) Mapping of soil properties at high resolution in Switzerland using boosted geoadditive models. Soil 3:191–210
Pachepsky Y, Rawls WJ (2004) Development of pedotransfer functions in soil hydrology. Elsevier, Amsterdam
Rab MA, Chandra S, Fisher PD, Robinson NJ, Kitching M, Aumann CD, Imhof M (2011) Modelling and prediction of soil water contents at field capacity and permanent wilting point of dryland cropping soils. Soil Res 49:389–407
Ramirez-Lopez L, Behrens T, Schmidt K, Stevens A, Demattê JAM, Scholten T (2013a) The spectrum-based learner: a new local approach for modeling soil vis–NIR spectra of complex datasets. Geoderma 195:268–279
Santana FB de, Souza AM de, Poppi RJ (2018) Visible and near infrared spectroscopy coupled to random forest to quantify some soil quality parameters. Spectrochim Acta Part A Mol Biomol Spectrosc 191:454–462
Tranter G, Minasny B, McBratney AB (2010) Estimating pedotransfer function prediction limits using fuzzy k-means with extragrades. Soil Sci Soc Am J 74:1967–1975
Van der Klooster E, Van Egmond FM, Sonneveld MPW (2011) Mapping soil clay contents in Dutch marine districts using gamma-ray spectrometry. Eur J Soil Sci 62:743–753
Viscarra-Rossel RA, Jeon YS, Odeh IOA, McBratney AB (2008) Using a legacy soil sample to develop a mid-IR spectral library. Soil Res 46:1–16
Wadoux AMJ-C, Brus DJ, Heuvelink GBM (2018) Accounting for non-stationary variance in geostatistical mapping of soil properties. Geoderma 324:138–147
Wehrens R, Mevik B-H (2007) The pls package: principal component and partial least squares regression in R. J Stat Softw 18:1–24
Williams PC, Thompson BN (1978) Influence of whole meal granularity on analysis of HRS wheat for protein and moisture by near infrared reflectance spectroscopy (NRS). Cereal Chem 55:1014–1037
Wold S, Johansson E, Cocchi M (1993) PLS: partial least squares projections to latent structures. In: 3D qsar in drug design: theory, methods and applications. Kluwer ESCOM Science, Dordrecht, pp 523–550
Wold S, Sjöström M, Eriksson L (2001) PLS-regression: a basic tool of chemometrics. Chemom Intell Lab Syst 58:109–130

Chapter 10
Spectral Transfer and Transformation

Measurement protocols for the same material often vary from laboratory to laboratory. Similarly, while the same spectrometer or sensor can be used between laboratories, the difference in terms of sensor or spectrometer manufacturer is likely to introduce additional variation in the recorded spectrum. These issues are relevant in soil spectroscopy, where there is a growing base of practitioners together with a growing number of available spectrometers to use. Despite a development of soil spectral libraries, the sharing of such libraries is not too common, which makes collaborative work difficult to coordinate. Importantly, soil spectral inference models calibrated at one laboratory are likely to be non-applicable to spectra collected from another laboratory.

There have been some efforts to compile large-scale libraries that are populated from locally derived sources and then building subsequent 'globally' calibrated models. To contribute to such projects, there is a need to adhere to certain minimum requirements and a measurement protocol for consistent measurement of spectra in the laboratory. However for the same soil sample, different spectrometers could produce different reflectance spectra because of the scanning protocol, lighting condition, instrumentation setting, sensors or white reference (Viscarra-Rossel et al. 2016). Similarly, external factors influencing the spectra, such as soil moisture, need to be accounted for in the modelling. This chapter provides examples to standardize the spectra scanned from different instruments, using either a standard spectrum as baseline or a reference spectrometer. An example on how to account for external factors affecting the spectra is also provided. Because this chapter makes use of model calibration described in Chap. 9, it is not considered as a pre-processing. Note, however, that the steps described in this chapter are most often used prior to or in combination with model calibration.

The set of packages used in this chapter are installed using the lines below. The book-associated `soilspec` package is also required; see Chap. 3 for information on its installation.

A. M. J.-C. Wadoux et al., *Soil Spectral Inference with R*, Progress in Soil Science,
https://doi.org/10.1007/978-3-030-64896-1_10

```
# specify all the packages used in the chapter and install them if they are not
    already
myPackages <- c("Cubist", "lattice", "prospectr", "RcppArmadillo",
               "pls", "MASS")

# define which packages are not installed in the current computer
notInstalled <- myPackages[!(myPackages %in% installed.packages()[ , "Package"])]

# install the missing packages
if(length(notInstalled)) install.packages(notInstalled)
```

10.1 Spectral Transfer Between Instruments Using a Standard Sample

Irrespective of whether one wants to contribute to data collation projects, measurement standardization methods are quite common. The idea of standardization entails an approach to minimize systematic effects between laboratories where an agreed-upon and well-known (species and concentration) material is used to align the readings of any method. This idea hails from the wet chemistry discipline (Willis 1972) and was drawn upon by Pimstein et al. (2011) who proposed a material internal soil standard (ISS) approach to be used for soil spectroscopic studies. Their work demonstrated that well-known and agreed-upon reference material that is measured under any set-up in any laboratory can be used to align one laboratory's spectral measurements to another's. Pimstein et al. (2011) described that an ideal ISS should be inexpensive, simple to use, easily delivered overseas, homogeneous, stable in space and time and useful for both radiometric and spectral calibration. The ISS also has to be as similar as possible to soil grain size (shape, size and nature), and if possible, it should have stable (and preferably chemically featureless) spectral performance across the entire spectral region. Ben Dor et al. (2015) proposed a couple of ISS that fulfil such criteria. These samples or the sites they were collected were characterized as bright, homogeneous sand dunes and situated along the coastline of Wylie Bay (WB) and Lucky Bay (LB) in southwestern Australia.

Ben Dor et al. (2015) describes the protocol for the collection of the spectra for each ISS to minimize potential measurement errors. Spectra were collected in both contact probe and dark box modes. When a request is made, the authors share the ISS material, together with the associated benchmark soil spectra, and suggest measurement protocol to perform an alignment. The user will then measure the ISS material with their own instrument following the suggested protocol for either contact probe or dark mode and then perform an alignment to match the benchmark spectra. The alignment is essentially the calculation of correction factors that when applied to newly collected soil spectra from the users' instrument will adjust the spectra as though it were collected from the instrument that was used to collect the benchmark spectra. Once the alignment is made, it will be possible to compare and share spectra from different laboratories that have also used the same ISS material and performed the alignment. This in turn also provides the opportunity to share associated soil spectral inference models.

The following exercise describes the key steps in the process for spectral standardization. The ISS material used for the standardization in this exercise is the LB sample. We will demonstrate the process used for standardizing spectra collected from two different vis-NIR instruments.

The necessary files for this section are provided in the book package `soilspec` in `data("datStand")` and described in Chap. 3.

```
require(soilspec)
# load the package data
data("datStand")

# show the content of the data
names(datStand)
```

```
## [1] "CSIRO.s"           "ISS_subordinates" "asd.soil"          "se.soil"
```

Lucky Bay ISS

The ISS are provided by Ben Dor et al. (2015). The benchmark spectra (as measured by contact probe) for both samples are found in the file `datStand$CSIRO.s`. This object is called `CSIROs`. The benchmark LB spectra is shown in Fig. 10.1.

```
# take data for this section
CSIROs <- datStand$CSIRO.s

# take the wavelength
wavelength <- as.numeric(colnames(CSIROs))

#Make plot
matplot(wavelength, t(CSIROs),
     type= "l",
     ylab = "Reflectance",
     xlab = "Wavelength /nm",
     ylim = c(0.2, 1.2),
     lty = 1)

# add a legend
legend("topleft",
       legend = c("Lucky Bay", "Wilie Bay"),
       lty = c(1, 1),
       col = c("black", "red"))
```

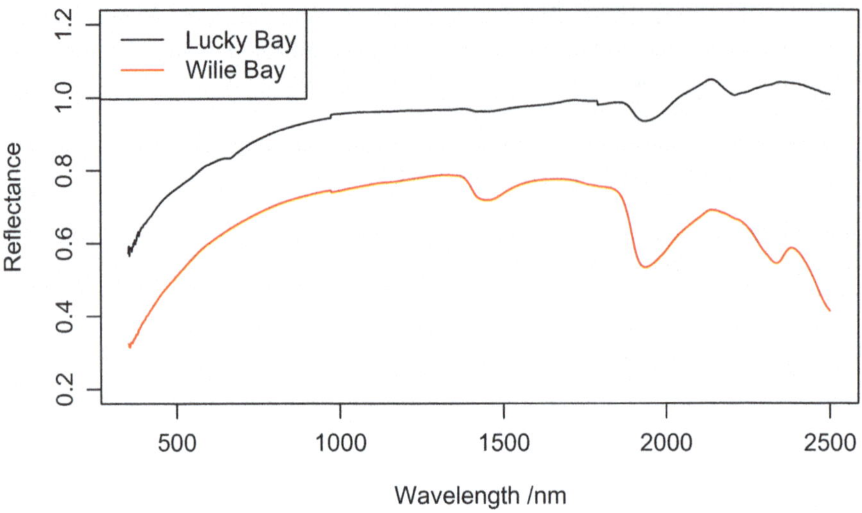

Fig. 10.1 Benchmark Lucky Bay (black line) and Wylie Bay (red line) spectra as measured by contact probe

Spectral collection The Lucky Bay ISS was measured with two different vis-NIR instruments:

1. Analytical Spectral Devices (ASD) AgricSpec^{TM} spectroradiometer.
2. Spectral Evolution PSR+ 3500 field portable spectroradiometer.

Both instruments collect high-resolution vis-NIR data and output the data to 1 nm resolution with spectral range of 350–2500 nm. Ben Dor et al. (2015) suggested a protocol for collecting the spectra of the ISS material. We followed the protocol for contact probe measurement. The spectra collected from each instrument are saved to the file `datStand$ISS_subordinates`. Figure 10.2 shows the benchmark spectra (black), with the corresponding spectra collected from the ASD (red) and Spectral Evolution instruments (blue).

```
# take the data
ISSsubordinates <- datStand$ISS_subordinates

# plot the benchmark spectra
plot(wavelength, CSIROs[1,],
     type= "l",
     ylab="Reflectance",
     ylim = c(0.5, 1.2),
     xlab = "Wavelength /nm")

# add line for ASD instrument
lines(wavelength, ISSsubordinates[1,2:ncol(ISSsubordinates)],
      col="red")

# add line for Spectral Evolution instrument
lines(wavelength, ISSsubordinates[2,2:ncol(ISSsubordinates)],
      col="blue")
```

```
# add a legend
legend("topleft",
       legend = c("Lucky Bay", "ASD", "Spectral Evolution"),
       lty = c(1, 1),
       col = c("black", "red", "blue"))
```

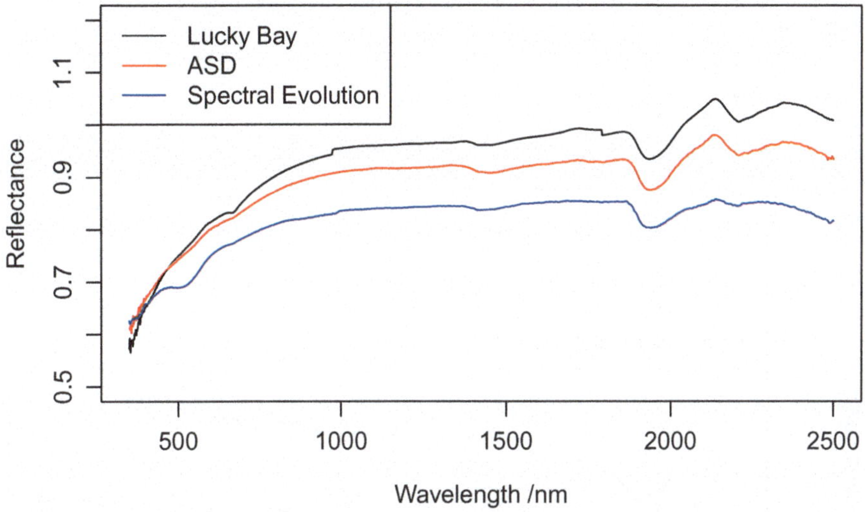

Fig. 10.2 Benchmark Lucky Bay spectra (black), with the corresponding spectra collected from the ASD (red) and Spectral Evolution (blue) instruments

Spectral alignment Aligning the spectral reading to the benchmark spectra is an important step. It is based on Pimstein et al. (2011) and described in Ben Dor et al. (2015). The first step is the estimation of correction factors that will align the collected spectra with the benchmark spectra. This is done using the following equation:

$$\mathbf{c}_\lambda = 1 - ((\mathbf{x}_{\text{sub}} - \mathbf{x}_{\text{ref}})/\mathbf{x}_{\text{sub}}), \tag{10.1}$$

where $\mathbf{c}_\lambda$ is the spectral correction factor vector; $\mathbf{x}_{\text{sub}}$ is the reflectance of the *subordinate* reference, or the ISS spectra from the users' instrument; and $\mathbf{x}_{\text{ref}}$ is the reflectance of the *reference*, or the benchmark ISS spectra (Fig. 10.3).

```
# correction factors for ASD instrument
xSub1 <- ISSsubordinates[1,2:ncol(ISSsubordinates)]
ASDcf <- 1- ((xSub1 - CSIROs[1,])/xSub1)

# correction factors for Spectral Evolution instrument
xSub2 <- ISSsubordinates[2,2:ncol(ISSsubordinates)]
seCf <- 1- ((xSub2 - CSIROs[1,])/xSub2)
```

```
# plot correction factors for ASD instrument
plot(wavelength, ASDcf,
     type = "l",
     ylab = "Correction factor",
     xlab = "Wavelength /nm",
     col = "red",
     ylim = c(0.9,1.3))

# plot correction factors for Spectral Evolution instrument
lines(wavelength, seCf,
      col = "blue")

# add a legend
legend("topleft",
       legend = c("ASD", "Spectral Evolution"),
       lty = c(1, 1),
       col = c("red", "blue"))
```

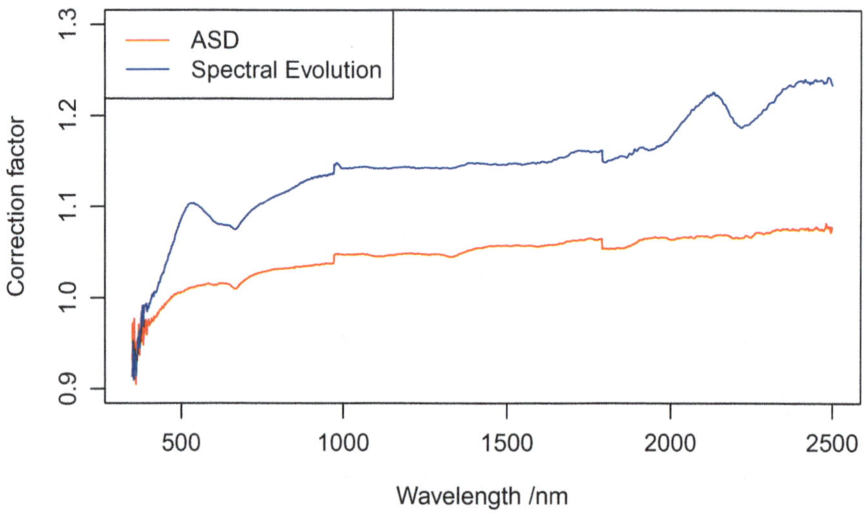

Fig. 10.3 Correction factors for the ASD (red) and Spectral Evolution (blue) instruments

The correction factors are then used to align newly collected spectra from each instrument to the benchmark using the following equation:

$$\mathbf{x}'_{\text{sub}} = \mathbf{x}_{\text{sub}} \times \mathbf{c}_{\lambda}, \tag{10.2}$$

where $\mathbf{x}'_{\text{sub}}$ is the corrected sample reflectance and $\mathbf{x}_{\text{sub}}$ is the original sample reflectance. We can use this equation to align the collected ISS spectra to the benchmark spectra to assess whether the correction was a success.

```
# align ASD spectra
asd.a <- xSub1 * ASDcf

# align Spectral Evolution spectra
se.a <- xSub2 * seCf

# difference between benchmark and ASD spectra
sum(abs(CSIROs[1,] - asd.a))
```

```
## [1] 7.693846e-14
```

```
# difference between benchmark and Spectra Evolution spectra
sum(abs(CSIROs[1,] - se.a))
```

```
## [1] 7.072121e-14
```

Soil spectra Two soil samples were scanned by both instruments. These are called `HVb21` and `HVd13`. The ASD scanned soils are in the file `datStand$asd.soil`, while those measured with the Spectral Evolution instrument are in `datStand$se.soil` which we will call `asdSoil` and `seSoil`, respectively (Figs. 10.4 and 10.5).

```
# rename the datasets
asdSoil <- datStand$asd.soil
seSoil <- datStand$se.soil

# plot spectrum of first sample scanned by ASD
plot(wavelength, asdSoil[1,2:ncol(asdSoil)],
     type = "l",
     ylim = c(0, 0.5),
     col = "red",
     ylab ="Reflectance",
     xlab = "Wavelength /nm")


# spectral Evolution measured
lines(wavelength, seSoil[1,2:ncol(seSoil)],
      col = "blue")

# add a legend
legend("topleft",
       legend = c("ASD", "Spectral Evolution"),
       lty = c(1, 1),
       col = c("red", "blue"))
```

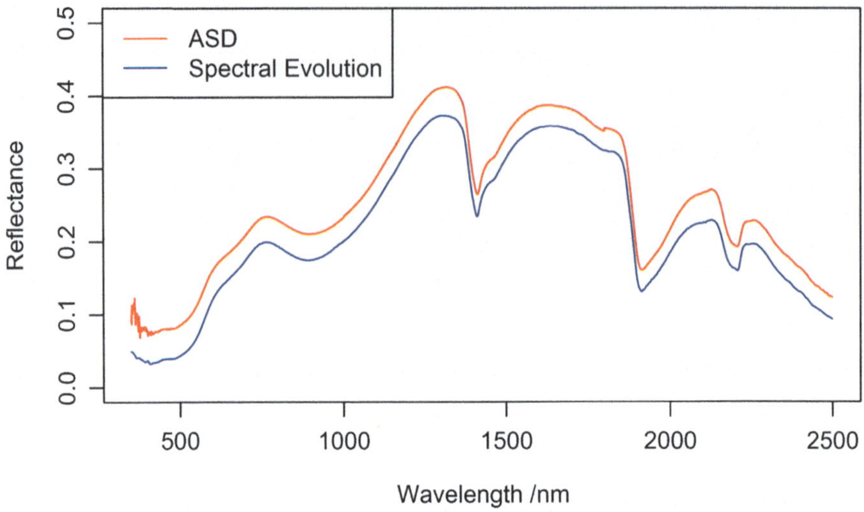

Fig. 10.4 Soil sample `HVb21` reflectance spectra scanned by the ASD (red) or Spectral Evolution (blue) instruments

```
# plot spectrum of first sample scanned by ASD
plot(wavelength, asdSoil[2,2:ncol(asdSoil)],
     type = "l",
     ylim = c(0, 0.6),
     col = "red",
     ylab = "Reflectance",
     xlab = "Wavelength /nm")

# spectral Evolution measured
lines(wavelength, seSoil[2,2:ncol(seSoil)],
      col="blue")

# add a legend
legend("topleft",
       legend = c("ASD", "Spectral Evolution"),
       lty = c(1, 1),
       col = c("red", "blue"))
```

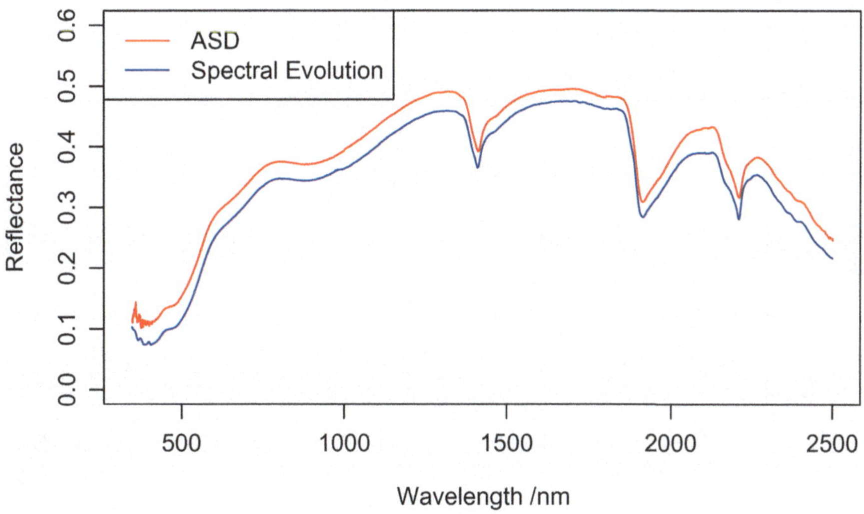

Fig. 10.5 Soil sample `HVd13` reflectance spectra scanned by the ASD (red) or Spectral Evolution (blue) instruments

Then we perform the alignments using the correction factors that were derived for each instrument before (Figs. 10.6 and 10.7).

```
# alignment sample 1 (HVb21)
alignSam1Asd <- asdSoil[1,2:ncol(asdSoil)] * ASDcf
alignSam1Se <- seSoil[1,2:ncol(seSoil)] * seCf

# plotting the aligned ASD spectrum
plot(wavelength, alignSam1Asd,
     type = "l",
     col = "red",
     ylim = c(0, 0.5),
     ylab = "Reflectance",
     xlab = "Wavelength /nm")

# adding line for the aligned Spectra Evolution
    spectrum
lines(wavelength, alignSam1Se,
      col = "blue")

# add a legend
legend("topleft",
       legend = c("ASD", "Spectral Evolution"),
       lty = c(1, 1),
       col = c("red", "blue"))
```

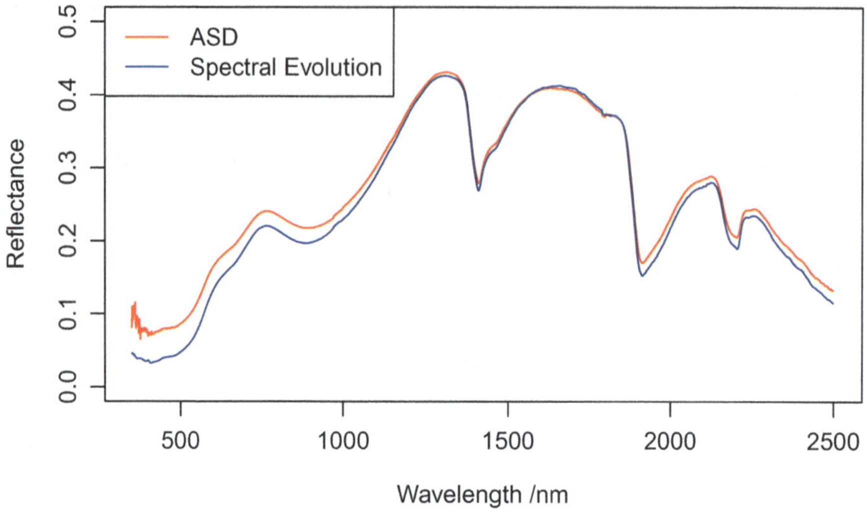

Fig. 10.6 Aligned ASD (red) and Spectral Evolution (blue) spectra for the soil sample `HVb21`

```
# alignment sample 2 (HVd13)
alignSam2Asd <- asdSoil[2,2:ncol(asdSoil)] * ASDcf
alignSam2Se <- seSoil[2,2:ncol(seSoil)] * seCf

# plotting the aligned ASD spectrum
plot(wavelength, alignSam2Asd,
     type = "l",
     ylim = c(0, 0.6),
     col = "red",
     ylab = "Reflectance",
     xlab = "Wavelength /nm")

# adding line for the aligned Spectra Evolution
    spectrum
lines(wavelength, alignSam2Se,
      col = "blue")

# add a legend
legend("topleft",
       legend = c("ASD", "Spectral Evolution"),
       lty = c(1, 1),
       col = c("red", "blue"))
```

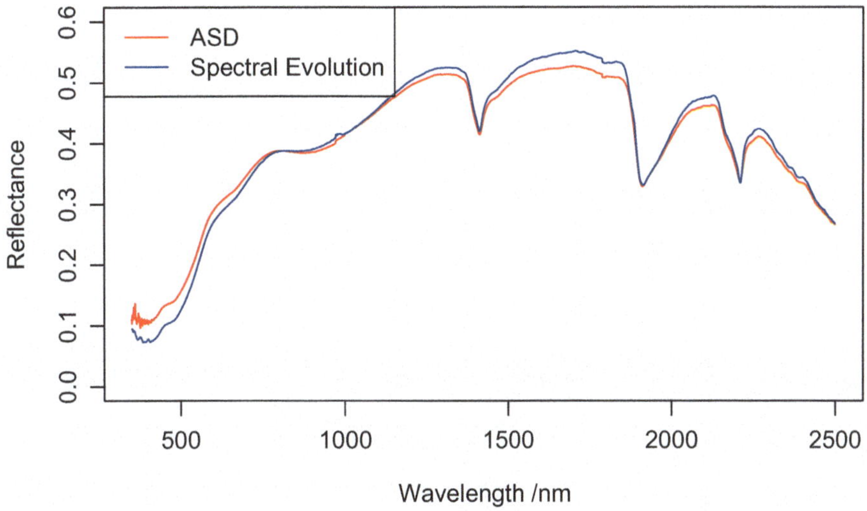

Fig. 10.7 Aligned ASD (red) and Spectral Evolution (blue) spectra for the soil sample HVd13

After the correction or alignment is done, one can exchange, compare or share spectra data with other users who used the exact ISS.

10.2 Direct Standardization

The approach outlined in the previous section used one or several standard samples for spectra standardization. It mostly corrects for the shift in the reflectance and does not deal with spectral narrowing or broadening in different instruments. A more general approach in standardization of spectra is called the multivariate standardization. The direct standardization (DS) method was developed for this purpose (Wang et al. 1991) but can also be used to correct for other external factors influencing the spectra.

In this example, we will illustrate the use of direct standardization by assuming that `spectra1snv` and `spectra0snv` come from two different instruments, where `spectra0Snv` are the spectra from the reference instrument and `spectra1Snv` are the spectra from the subordinate instrument which need to be standardized. Both `spectra0snv` and `spectra1snv` are pre-processed spectra of two sets of spectra found in the `datEPO` dataset of the `soilspec` package. In practice, a small set of samples are scanned in both instruments, while we have a larger number of samples to standardize. In this example, we assume that 20 samples have been scanned with both instruments and that the we want to apply the transfer function to a larger number of samples, which we call `subordinateSpec`.

```
# load the required packages
require(soilspec)
require(prospectr)

# load the EPO dataset from the soilspec package
data("datEPO")

# take the data from EPO, as an example between subordinate and reference spectra
spectra0 <- datEPO$spectra0
spectra1 <- datEPO$spectra1

# apply Savitzky-Golay filter
spectra0_sg <- savitzkyGolay(spectra0, w = 11, p = 2, m = 0)
spectra1_sg <- savitzkyGolay(spectra1, w = 11, p = 2, m = 0)

# resample the spectra
new.wavs <- seq(500, 2450, by = 1)
spectra0_rs <- prospectr::resample(spectra0_sg,
                                   wav = as.numeric(colnames(spectra0_sg)),
                                   new.wav = new.wavs,
                                   interpol = "linear")
spectra1_rs <- prospectr::resample(spectra1_sg,
                                   wav = as.numeric(colnames(spectra1_sg)),
                                   new.wav = new.wavs,
                                   interpol = "linear")

# apply a standard normal variate transformation for baseline correction
spectra0Snv <- standardNormalVariate(spectra0_rs)
spectra1Snv <- standardNormalVariate(spectra1_rs)
# define the samples scanned in both instruments
referenceSpecSub <- spectra0Snv[1:20,]
subordinateSpecSub <- spectra1Snv[1:20,]

# the unstandardized spectra are called subordinateSpec
referenceSpec <- spectra0Snv
subordinateSpec <- spectra1Snv

# plot the reference and subordinate spectra used to build the transfer function
matplot(colnames(referenceSpecSub), t(referenceSpecSub),
        main = " ",
        xlab = "Wavelength /nm",
        ylab = "Absorbance",
        type = "l",
        col = "blue",
        lty = 1)

# add the subordinate spectra to the plot
matlines(colnames(subordinateSpecSub), t(subordinateSpecSub),
         col="pink")

# add a legend
legend("topright",
       legend = c("reference", "subordinate"),
       lty = c(1, 1),
       col = c("blue", "pink"))
```

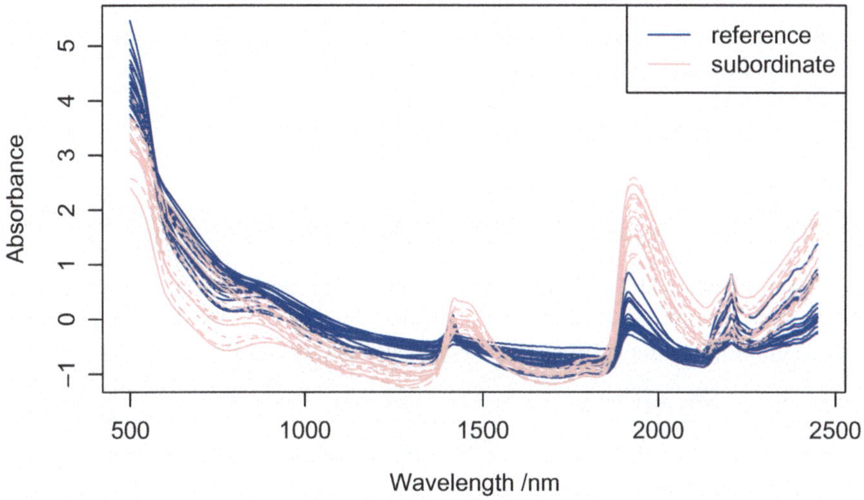

Fig. 10.8 Reference and subordinate spectra used to build the transfer function

Figure 10.8 shows that spectra from the subordinate instrument have a higher reflectance in the region between 1200 and 1500 nm and between 1850 and 2100 nm.

The basis of the direct standardization transformation is based on a simple matrix operation. We can write the subordinate soil spectra $\mathbf{X}_{\text{sub}}$ (`spectra1Snv`) as a function of the reference instrument spectra $\mathbf{X}_{\text{ref}}$ (`spectra0Snv`):

$$\mathbf{X}_{\text{ref}} = \mathbf{X}_{\text{sub}}\mathbf{P} + \mathbf{e}, \tag{10.3}$$

where $\mathbf{e}$ is an independent error vector and $\mathbf{P}$ is a transformation matrix, which can be obtained by taking the generalized inverse of $\mathbf{X}_{\text{sub}}$:

$$\mathbf{P} = \mathbf{X}^{g}_{\text{sub}}\mathbf{X}_{\text{ref}}, \tag{10.4}$$

where $\mathbf{X}^{g}_{\text{sub}}$ is the generalized inverse of $\mathbf{X}_{\text{sub}}$. We can implement it in R using the following lines. The function takes as argument `Xs` which is the subordinate spectra and `Xm` which is the reference spectra. The function output is the matrix transformation `P`.

```
# define the DS function
DS <- function(reference, subordinate){

  # load the required package
  library(MASS)

  # subordinateI is the generalized inverse of spectra
      Xs
  subordinateI = ginv(subordinate)
```

```
  # T is the transformation matrix from Xs to Xm
  P = subordinateI%*%reference
  return(P)
}
```

We can now apply the DS function to the datasets.

```
P <- DS(reference = referenceSpecSub, subordinate = subordinateSpecSub)
```

Matrix **P**, denoted P, is the transfer matrix. The next step is to transfer the subordinate spectra so it is equivalent to the reference spectra. We do so by applying a matrix multiplication between the subordinate spectra and **P**.

```
Dspec <- as.matrix(subordinateSpec) %*% P
```

We can now visualize the spectra from the subordinate device corrected by the reference instrument using the matrix **P**.

```
# plot the reference and standardized subordinate spectra via DS
matplot(colnames(referenceSpec), t(referenceSpec),
        xlab = "Wavelength /nm",
        ylab = "Absorbance",
        type = "l",
        col = "blue",
        lty = 1)

# add the corrected subordinate spectra to the plot
matlines(colnames(Dspec), t(Dspec), col = "red")

# add a legend
legend("topright",
       legend = c("reference", "corrected subordinate"),
       lty = c(1, 1),
       col = c("blue", "red"))
```

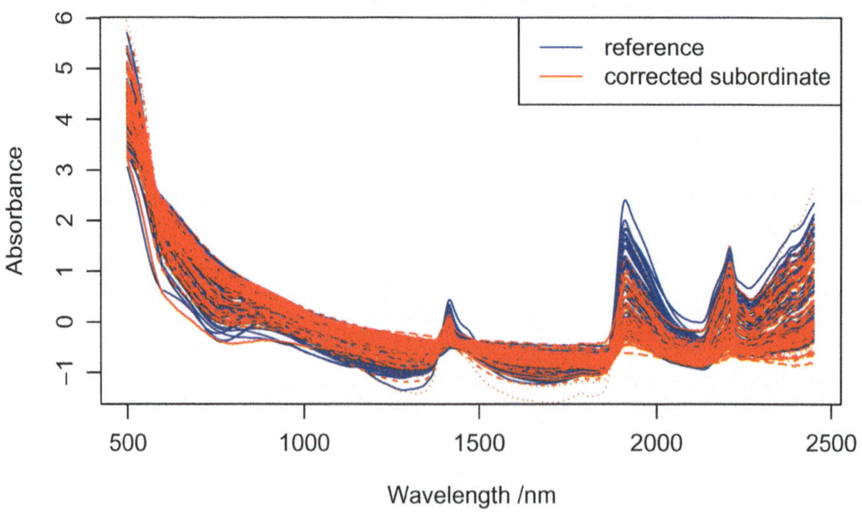

Fig. 10.9 Spectra from the reference (blue) and corrected spectra (red) from the subordinate instrument

A visual assessment of Fig. 10.9 shows that the transformed spectra mimic the spectra from the reference instrument.

The DS approach is a simple method to transfer spectra between instruments; however, the transformation matrix **P** is often an over-determined system, which means that the number of parameters is larger than the number of observations. In this example, the size of matrix **P** is 1951 where we only have 20 observations, i.e. 20×1951 values. This also means that to transform a spectrum at a particular wavelength, the information of the whole spectrum is used. In the next section, the piecewise direct standardization is used to perform a local transformation, where only the information from a local window is used to transform the spectrum at a particular wavelength.

10.3 Piecewise Direct Standardization

Piecewise direct standardization (PDS) tackles the over-determined system of the direct standardization transformation, i.e. the transformation matrix in direct standardization uses the whole spectrum to standardize a particular wavelength. PDS, proposed by Wang et al. (1991), makes the transformation of the value at a particular wavelength using only its neighbouring values defined by a moving window on the spectrum, which restricts **P** to be a banded matrix.

PDS only calculates transformation coefficients along the diagonal of **P**. For example, for wavelength i, a transfer function is determined by linear regression using the neighbouring values as independent variables in the model. In other words, the transfer function relates reflectance or absorbance of spectra from the subordinate instrument at wavelength i, $\mathbf{x}_{\text{ref},i}$ with reflectance values of spectra from reference instrument surrounding i with window size w, $\mathbf{X}_{\text{sub},i} : (x_i - w, \ldots, x_i, \ldots, x_i + w)$:

$$\mathbf{x}_{\text{ref},i} = \mathbf{X}_{\text{sub},i}\mathbf{b}_i + a + \mathbf{e}, \tag{10.5}$$

where $\mathbf{b}_i$ is the vector of linear regression coefficients at wavelength i and a is the intercept of the model. The linear model is fitted for each wavelength separately, and coefficients **b** are stored in matrix **P**:

$$\mathbf{P} = \text{diag}(\mathbf{b}_1^T, \mathbf{b}_2^T, \ldots, \mathbf{b}_i^T, \ldots, \mathbf{b}_b^T), \tag{10.6}$$

where b is the number of wavelengths in the spectra. Elements of most of the off-diagonal elements are set to zero. To calculate the standardized subordinate spectra $\tilde{\mathbf{X}}_{\text{sub}}$, we can use a matrix operation:

$$\tilde{\mathbf{X}}_{\text{sub}} = \mathbf{X}_{\text{sub}}\mathbf{P}. \tag{10.7}$$

Since an intercept is also calculated to account for bias of the spectra, the intercept needs to be applied to each of the transformed spectra $\tilde{\mathbf{x}}_{\text{sub}}$:

$$\tilde{\mathbf{x}}_{\text{sub}}^{T} = \tilde{\mathbf{x}}_{\text{sub}} + \mathbf{a}^{T}, \tag{10.8}$$

where $\tilde{\mathbf{x}}_{\text{sub}}^{T}$ is the bias-adjusted PDS transformed spectrum and $\mathbf{a}$ is the vector of intercept values.

The PDS function can be scripted as follows (courtesy of Guillaume Hans, FPInnovations, Canada).

```
# define the PDS function
PDS <- function(reference, subordinate, ws, ncomp){

  # load the required package
  require(pls)

  # define the variables
  nc <- ncol(reference)
  windowSize <- (ws -1)/2
  k <- windowSize-1

  # create an empty Transformation matrix P and Intercep matrix
  P <- matrix(0, nrow = nc, ncol = nc-(2*windowSize)+2)
  Intercep <- matrix(0, nrow = nc, ncol = 1)
  np <- windowSize*2 # no parameters of PLSR
  loop <- windowSize:(nc-k)

  for(i in loop){

    #PLS regression:
   pls.mod <- plsr(reference[,i] ~ as.matrix(subordinate[,(i-k):(i+k)]), ncomp)

    #extract PLSR coefficients
    coefPLS <- as.numeric(coef(pls.mod, ncomp, intercept = TRUE))

    # Save intercept
    Intercep[i] <- coefPLS[1]

    # Save coefficients to matrix P
    P[(i-k):(i+k), i-k] <- t(coefPLS[2:np])

  }

  # account for the edges by adding 0 values
  P <- data.frame(matrix(0,nrow = ncol(reference), ncol = k), P,
                  matrix(0, nrow = ncol(reference), ncol = k))

  colnames(P) <- colnames(reference)
  PDSpar <-list(P = P, Intercep = Intercep)

  return(PDSpar)
}
```

The PDS function takes as argument the spectra from subordinate and reference instruments, the window size (must be an odd number) and the number of components. A partial least square (PLS) model (see Sect. 9.2) is used to relate the spectrum at wavelength i of the subordinate instrument to the bands of the reference instrument, for a given window size.

The following shows how the PDS is applied to the example we described previously, using `referenceSpecSub` and `subordinateSpecSub` spectra presented in Fig. 10.8:

```
# define the window size (must be odd number)
##note that here the spectra is resampled at every 5nm
## so that a window size of 3 is equal to 15nm
ws = 3

# apply the PDS function
PDSpar <- PDS(reference = referenceSpecSub,
              subordinate = subordinateSpecSub,
              ws,
              ncomp = 1)
```

The `PDSpar` variable contains the matrix `P` and matrix with a single column containing the intercept values of the PLS model. We can now apply Equations (10.7) and (10.8) to the subordinate spectra.

```
# multiply spectra with transformation matrix
tSpec <- subordinateSpec %*% as.matrix(PDSpar$P)

# add to each row the intercept values
for (i in 1:nrow(tSpec)){
tSpec[i,] <- tSpec[i,] + as.numeric(t(PDSpar$Intercep))
}
```

The corrected subordinate spectra can then be plotted.

```
# plot the reference and subordinate spectra
matplot(as.numeric(colnames(referenceSpec)), t(referenceSpec),
        col = "blue",
        lty = 1,
        xlab = "Wavelength /nm",
        ylab = "Absorbance",
        type = "l")

# add the corrected subordinate to the plot
matlines(as.numeric(colnames(tSpec)), t(tSpec),
         col = "red")

# add a legend
legend("topright",
       legend = c("reference", "corrected subordinate"),
       lty = c(1, 1),
       col = c("blue", "red"))
```

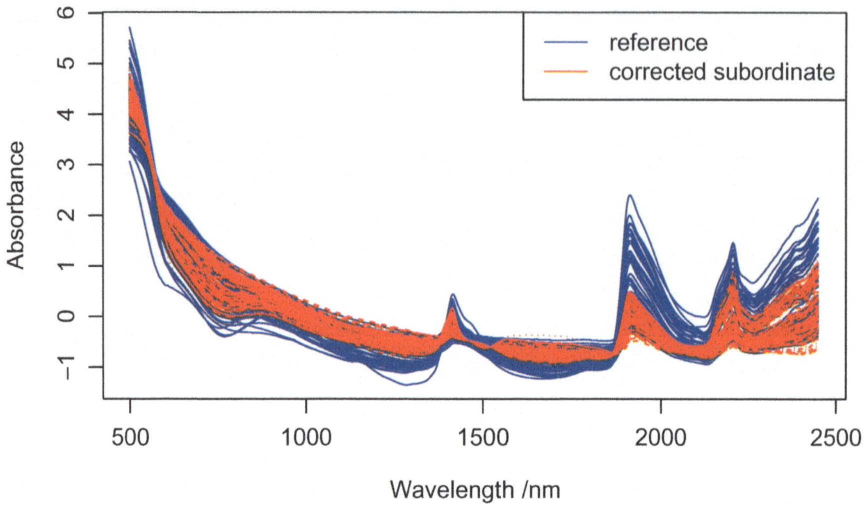

Fig. 10.10 Reference (blue) and corrected (red) subordinate spectra. The corrected spectra in red has been corrected using piecewise direct standardization

Figure 10.10 shows both the spectra from the reference instrument and the corrected by PDS spectra from the subordinate instrument. Note that the edges should be removed because of the application of a moving window operation to the spectra. The use of a moving window on the spectra is discussed previously in Chap. 5.

10.4 Removing External Effects, such as Soil Moisture (EPO)

Spectra in the infrared range are sensitive to external environmental conditions, such as soil moisture. The absorbance spectrum of a soil increases with increasing moisture content. In the laboratory, soil is usually scanned under standard air-dried conditions, but in the field it is difficult to control the water content. Although many studies have successfully used spectra collected in field conditions to calibrate against measured soil properties (e.g. SOC), the variation of soil moisture content can really have an important impact on the prediction of soil properties.

This section presents the external parameter orthogonalization (EPO) method to remove the moisture effect from the spectral calibration. The EPO algorithm projects all soil spectra orthogonal to the space of unwanted variation (i.e. moisture), and thus the variations due to soil moisture can be effectively removed.

```
# load the required packages
require(prospectr)
require(soilspec)

# load the package data
data("datsoilspc")

# convert reflectance to absorbance
spectraA <- log(1/datsoilspc$spc)

# embed the soil property and the spectra in one single table
datsoilspc$spcA <- spectraA

# apply a Savitzky-Golay filter
datsoilspc$spcASg <- savitzkyGolay(datsoilspc$spcA, w = 11, p = 2, m = 0)


# apply some smoothing to the spectra
old.wavs <- as.numeric(colnames(datsoilspc$spcASg))
new.wavs <- seq(500, 2450, by = 5)
datsoilspc$spcARs <- prospectr::resample(datsoilspc$spcASg,
                                         wav = old.wavs,
                                         new.wav = new.wavs,
                                         interpol = "linear")

# apply a standard normal variate transformation for baseline correction
datsoilspc$spcASnv <- standardNormalVariate(datsoilspc$spcARs)
```

In this example, from Minasny et al. (2011), we have 100 soil samples (a subset of the larger dataset) under 3 different moisture conditions. So we have three sets of spectra data with different moisture contents.

- `spectra0` is absorbance spectra for soil under air-dried condition (average moisture 5%).
- `spectra1` is absorbance spectra for soil after being wetted (average moisture 12%).
- `spectra2` is absorbance spectra for wetted soil after being air-dried for 1 day (average moisture 9%).

They are comprised in the dataset called `datEPO` from the book package `soilspec`. Let us first load the data and rename the objects.

```
# import the data
data("datEPO")

# show names
names(datEPO)
```

```
## [1] "soilC"    "spectra0" "spectra1" "spectra2"
```

```
# extract the objects from datEPO
soilC <- datEPO$soilC
spectra0 <- datEPO$spectra0
spectra1 <- datEPO$spectra1
spectra2 <- datEPO$spectra2
```

We can apply some pre-processing to the spectra the same way as for the `datsoilspc` spectra.

```
# apply Savitzky-Golay filter
spectra0_sg <- savitzkyGolay(spectra0, w = 11, p = 2, m = 0)
spectra1_sg <- savitzkyGolay(spectra1, w = 11, p = 2, m = 0)
spectra2_sg <- savitzkyGolay(spectra2, w = 11, p = 2, m = 0)

# make some preprocessing to the spectra
spectra0_rs <- prospectr::resample(spectra0_sg,
                                   wav = old.wavs,
                                   new.wav = new.wavs,
                                   interpol = "linear")
spectra1_rs <- prospectr::resample(spectra1_sg,
                                   wav = old.wavs,
                                   new.wav = new.wavs,
                                   interpol = "linear")
spectra2_rs <- prospectr::resample(spectra2_sg,
                                   wav = old.wavs,
                                   new.wav = new.wavs,
                                   interpol = "linear")

# apply a standard normal variate transformation for baseline correction
spectra0Snv <- standardNormalVariate(spectra0_rs)
spectra1Snv <- standardNormalVariate(spectra1_rs)
spectra2Snv <- standardNormalVariate(spectra2_rs)
```

As an example, we can see the spectra for sample 1 under three (`spectra0Snv`, `spectra1Snv` and `spectra2Snv`) different moisture conditions.

```
# plot the spectra0
plot(colnames(spectra0Snv), spectra0Snv[1, ],
     type = "l",
     xlab = "Wavelength /nm",
     ylab = "Absorbance",
     col = "blue")

# add line for spectra2
lines(colnames(spectra2Snv), spectra2Snv[1, ],
      col = "red")

# add line for spectra1
lines(colnames(spectra1Snv), spectra1Snv[1, ],
      col = "green")

# add a legend
legend("topright",
   legend = c("average moisture 5%", "average moisture 9%", "average moisture 12%"),
       col = c("blue","red" ,"green"),
       lty = 1,
       cex = 1)
```

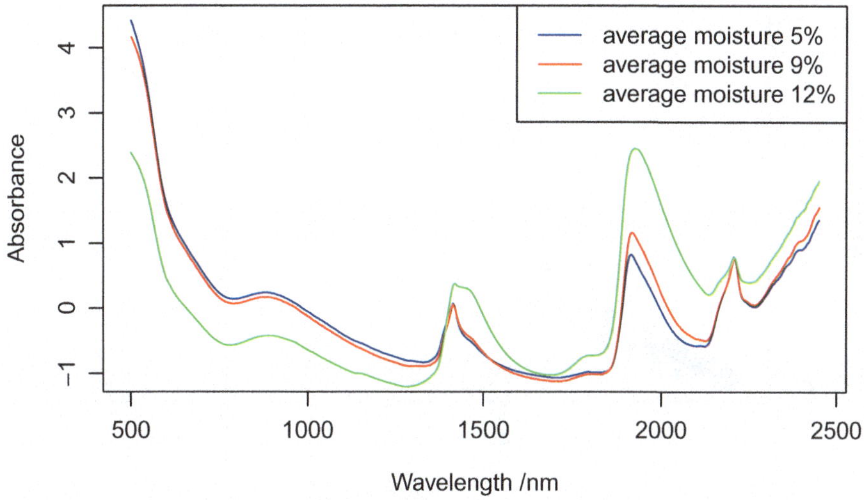

Fig. 10.11 Three example spectra scanned with different levels of soil moisture for low (5%, blue line), moderate (9%, red line) and high (12%, green line) soil moisture content

Figure 10.11 shows the influence of moisture content on the spectra.

Usually, one makes a model based on the library (of dried samples) and then uses the model built from the library to predict soil properties from spectra collected in the field (with varying moisture content). We now want to investigate the effect of moisture content on the prediction. Note the moisture bands at wavelengths around 1400 and 1900 nm (Fig. 10.12).

```
# load required package
require(Cubist)

# compute the integer number corresponding to the 25% of the samples
nsamples2select <- round(x = nrow(datsoilspc$spcASnv) * 0.75, digits = 0)

# separate between calibration and validation
isrow <- sample(1:nrow(datsoilspc$spcASnv), size = nsamples2select)

datsoilspcC <- datsoilspc[isrow,]
datsoilspcV <- datsoilspc[-isrow,]

# generate a Cubist model
soilcCubistModel <- cubist(x = datsoilspcC$spcASnv, y = datsoilspcC$TotalCarbon)

# predict on the calibration dataset
soilvCubistPred <- predict(soilcCubistModel, datsoilspcV$spcASnv)

# plot the predicted calibration data
plot(datsoilspcV$TotalCarbon, soilvCubistPred,
     xlab = "Observed",
     ylab = "Predicted",
     xlim = c(0, 12),
     ylim = c(0, 12),
     pch = 16)
abline(0, 1)
```

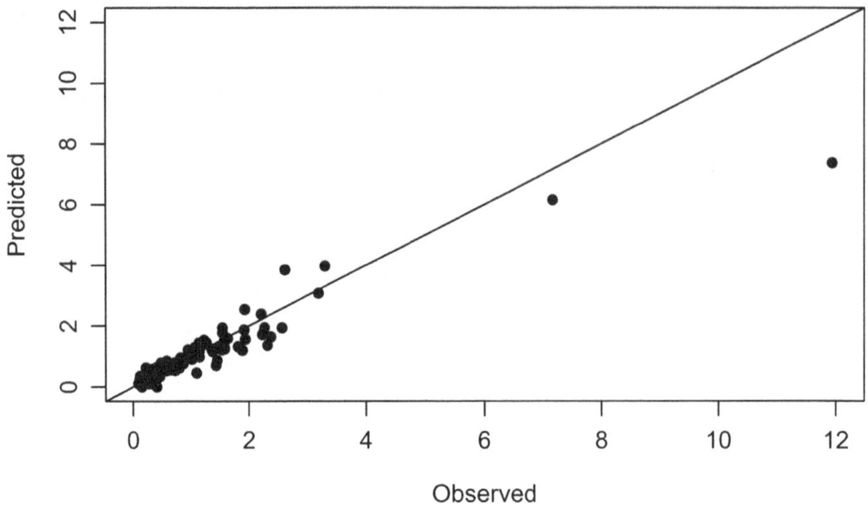

Fig. 10.12 Scatterplot of observed against predicted values of total carbon. Predictions are made by the cubist model on the calibration dataset for the example

We evaluate the prediction using the `eval` function from the book-associated `soilspec` package.

```
soilspec::eval(datsoilspcV$TotalCarbon, soilvCubistPred, obj = "quant")
```

```
##         ME RMSE   r2   R2 rhoC RPD RPIQ
## 1 -0.09 0.56 0.86 0.85 0.91 2.6 2.12
```

```
# predict the values from spectra at different moisture content
soilvCubistPredDry <- predict(soilcCubistModel, spectra0Snv)
soilvCubistPredWet <- predict(soilcCubistModel, spectra1Snv)
soilvCubistPredWet2 <- predict(soilcCubistModel, spectra2Snv)

par(mfrow = c(1, 3))

# plot the prediction on low soil moisture content spectra
plot(soilC$TotalC, soilvCubistPredDry,
     xlab = "Observed",
     ylab = "Predicted",
     main = "Moisture = 5%",
     xlim = c(0, 12),
     ylim = c(0, 12),
     pch = 16)
abline(0, 1)

# plot the prediction on moderate soil moisture content spectra
plot(soilC$TotalC, soilvCubistPredWet2,
     xlab = "Observed",
     ylab = "Predicted",
     main = "Moisture = 9%",
     xlim = c(0, 12),
     ylim = c(0, 12),
     pch = 16)
```

```
abline(0, 1)

# plot the prediction on high soil moisture content spectra
plot(soilC$TotalC, soilvCubistPredWet,
     xlab = "Observed",
     ylab = "Predicted",
     main = "Moisture = 12%",
     xlim = c(0, 12),
     ylim = c(0, 12),
     pch = 16)
abline(0, 1)
```

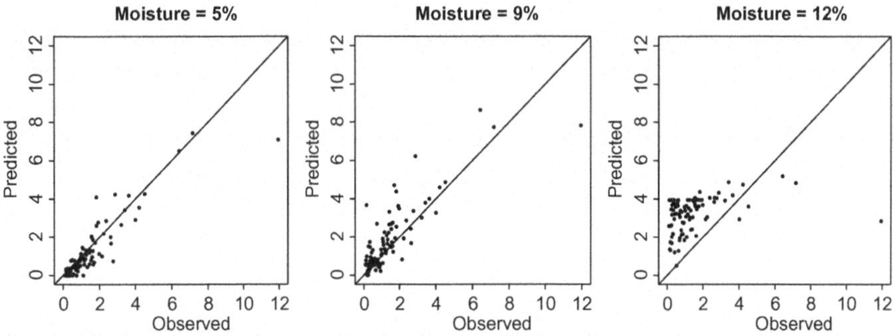

Fig. 10.13 Scatterplot of predicted against observed values of total carbon. Predictions are made using a cubist model on three varying levels of soil moisture from low (left) to high (right) soil moisture content

We also derive the accuracy measures (Sect. 9.1) between predicted and observed values of total carbon for the three datasets with varying levels of moisture.

```
# print accuracy measures for the three predictions
soilspec::eval(soilC$TotalC, soilvCubistPredDry, obj = "quant")
```

```
##      ME RMSE   r2   R2 rhoC  RPD RPIQ
## 1 -0.25 0.79 0.79 0.84 0.91 2.54 1.62
```

```
soilspec::eval(soilC$TotalC, soilvCubistPredWet2, obj = "quant")
```

```
##     ME RMSE   r2   R2 rhoC  RPD RPIQ
## 1 0.31 1.08 0.64 0.71 0.84 1.86 1.18
```

```
soilspec::eval(soilC$TotalC, soilvCubistPredWet, obj = "quant")
```

```
##     ME RMSE   r2    R2 rhoC  RPD RPIQ
## 1 1.81 2.61 0.14 -0.72 0.16 0.77 0.49
```

Figure 10.13 and the accuracy measures show that the prediction on the dry samples are accurate, whereas C content is underpredicted when the soil sample contains more moisture.

External Parameter Orthogonalization (EPO) An algorithm can be used to remove the 'unwanted' (external parameter influence) signal from the spectra. EPO (external parameter orthogonalization), developed by Roger et al. (2003), finds the areas in the spectra affected by external conditions (such as moisture) and projects the spectra orthogonal to this variation so that the unwanted components can be removed. This analysis is analogous to principal component analysis (PCA, Chap. 6), where the orthogonalization takes into account the variability which due to an external factor (i.e. moisture). The algorithm was developed by Roger et al. (2003) to remove the effect of temperature from the spectra for the prediction of measurement of the sugar content of fruit.

In EPO, it is assumed that the information contained in a spectrum can be decomposed into three components. The first component is the useful chemical spectral responses (also called spectra chromophores); the second component of the spectrum is caused by external factors independent of the first part. The last component is a residual.

In matrix form, the spectra **X** (size $n \times b$) can be written as:

$$\mathbf{X} = \mathbf{XP} + \mathbf{XQ} + \mathbf{R}, \tag{10.9}$$

where **P** is the projection matrix (size $b \times b$) of the useful part of the spectra, $\tilde{\mathbf{X}} = \mathbf{XP}$; **Q** is the projection matrix (size $b \times m$) of the unwanted (i.e. influenced by moisture) part of the spectra, $\mathbf{X}^T = \mathbf{XQ}$; and **R** is the residual matrix of size $n \times b$.

The aim of EPO is to obtain the useful part of the spectra:

$$\tilde{\mathbf{X}} = \mathbf{X}(\mathbf{I} - \mathbf{Q}), \tag{10.10}$$

while matrix **Q** is written as $\mathbf{Q} = \mathbf{GG}^T$. Matrix **G** is the uninformative part of the spectra that is orthogonal to the useful part of the spectra; Roger et al. (2003) suggested using the principal component of the difference spectra **D** as an approximation.

Here, we define **D** as the difference between the moist and air-dried (standard) conditions. In this example, we calculate **D** as the difference in spectra in the moist conditions (average w = 12%) and spectra at air-dried condition.

```
# difference matrix (spectra difference between wet and dry)
D = as.matrix(spectra0Snv - spectra1Snv)
```

The next step is to apply a PCA on **D** to extract the variation subspace. We do this by defining the EPO function in R, following Roger et al. (2003), and with two arguments. The first argument, `npc`, is the number of components to use. The second argument `D` is the difference matrix. The `epo()` function estimates **Q** by using a principal component analysis on $\mathbf{D}^T\mathbf{D}$ via a singular value decomposition (SVD) to obtain $\mathbf{USV}^T$, where **U** is an upper triangular matrix of size $n \times n$, **S** is a diagonal matrix of size $n \times b$ and **V** is a matrix of size $b \times b$. See Wall et al. (2003) for the relationship between PCA and SVD.

We need to define the number of factors npc to be used in EPO. This allows us to estimate **Q** (the uninformative part of the spectra) from the PC subset. Thus we can calculate the projection matrix as $\mathbf{P} = \mathbf{I} - \mathbf{Q}$.

The function epo() requires D and npc and returns P, that is, the projection matrix.

```
# define the EPO function
epo <- function(D, npc){
  # npc is the number of components to use
  # return: P: the projection matrix
  D <- as.matrix(D)
  n <- nrow(D)
  p <- ncol(D)

  dtd <- t(D) %*% (D)
  # singular value decomposition of the D(n x n) matrix
  s <- svd(dtd)
  # extract the no. factors
  ld <- s$v[, 1:npc]
  # projection matrix
  P <- diag(p) - ld %*% t(ld)

  return(P)
}
```

In this example, we decide to use three principal components. More information on how to select an optimal number of PC is provided in the next section.

```
# define number of principal components
npc <- 3
```

Using the epo() function, one can now obtain the projection matrix P.

```
# compute the projection matrix P:
P <- epo(D, npc)
```

The projection matrix P is visualized by the function myImagePlot of the book-associated soilspec package, which needs to be loaded first.

```
# load required package
require(soilspec)

# plot the projection matrix P
myImagePlot(P,
            zlim = c(-0.1,0.1))
```

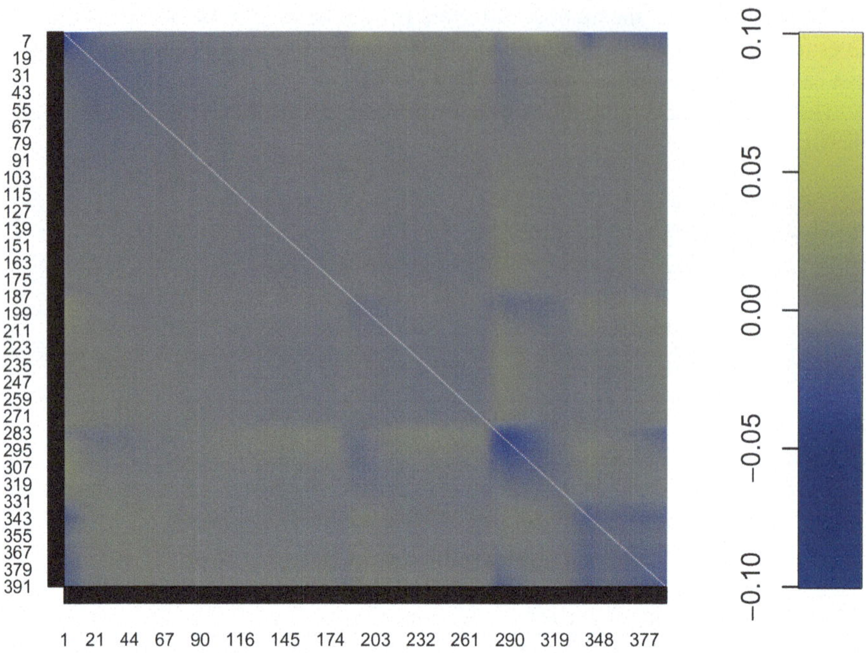

Fig. 10.14 Projection matrix computed using three components. The axes are the indices of the wavelength

The projection matrix displayed in Fig. 10.14 shows the wavelengths affected by moisture or external conditions. The areas with large positive or negative values indicate spectra regions that are going to be transformed. This corresponds to the parts of the spectra affected by moisture. The transformed spectra where the unwanted features have been removed are called **Z**: $\mathbf{Z} = \mathbf{XP}$. These transformed spectra are not affected by moisture and can be used in modelling.

We use matrix `P` to project the spectra.

```
# EPO projected spectra of spec0
Z0 <- as.matrix(spectra0Snv) %*% P
colnames(Z0) <- colnames(spectra0Snv)

# EPO projected spectra of spec1
Z1 <- as.matrix(spectra1Snv) %*% P
colnames(Z1) <- colnames(spectra1Snv)

# EPO projected spectra of spec2
Z2 <- as.matrix(spectra2Snv) %*% P
colnames(Z2) <- colnames(spectra2Snv)
```

If we plot the EPO transformed spectra **Z**, taking the spectrum of the first soil sample as example, it gives:

```
# plot corrected spectra0
plot(colnames(spectra0Snv), Z0[1,],
     type = "l",
     xlab = "Wavelength /nm",
     ylab = "Absorbance",
     col = "blue")

# add line for corrected spectra2
lines(colnames(spectra2Snv), Z2[1,],
      col = "red")

# add line for corrected spectra1
lines(colnames(spectra1Snv), Z1[1,],
      col = "green")

# add a legend
legend("topright",
    legend = c("average moisture 5%", "average moisture 9%", "average moisture 12%"),
       col = c("blue","red" ,"green"),
       lty = 1,
       cex = 1)
```

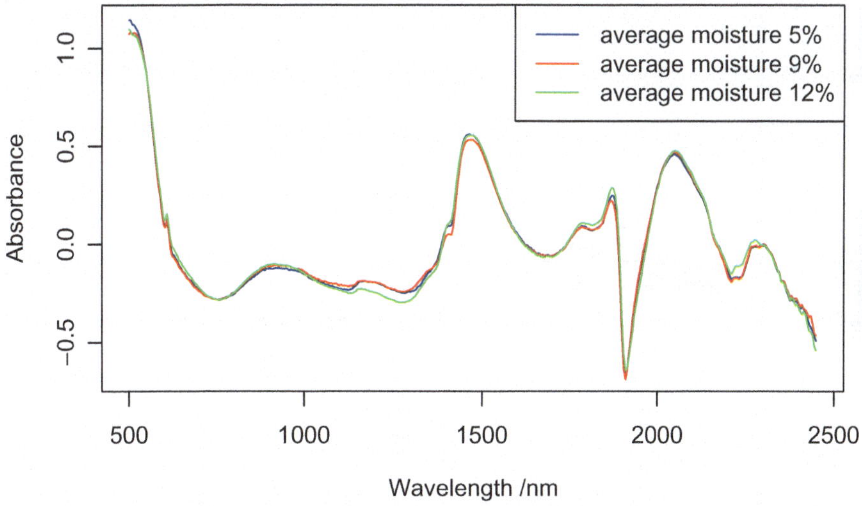

Fig. 10.15 Soil spectra corrected for soil moisture

Figure 10.15 shows that all three spectra are quite similar (they are from the same soil sample but at different moisture content), as opposed to what we see in Fig. 10.11. The next step consists in projecting the library spectra into transformed spectra using the `P` matrix and calibrating the new spectra to the required soil properties (Fig. 10.16).

```
# project the whole calibration spectra
specZ <- datsoilspc$spcASnv %*% P
colnames(specZ) <- colnames(datsoilspc$spcASnv)

specZC <- specZ[isrow,]
specZV <- specZ[-isrow,]

# make a Cubist model from EPO transformed dry spectra
epoCubistModel <- cubist(x = specZC, y = datsoilspcC$TotalCarbon)

# predict on the calibration dataset
soilvCubistPred <- predict(epoCubistModel, specZV)

# plot the predicted calibration data
plot(datsoilspcV$TotalCarbon, soilvCubistPred,
     xlab = "Observed",
     ylab = "Predicted",
     xlim = c(0, 12),
     ylim = c(0, 12),
     pch = 16)
abline(0, 1)
```

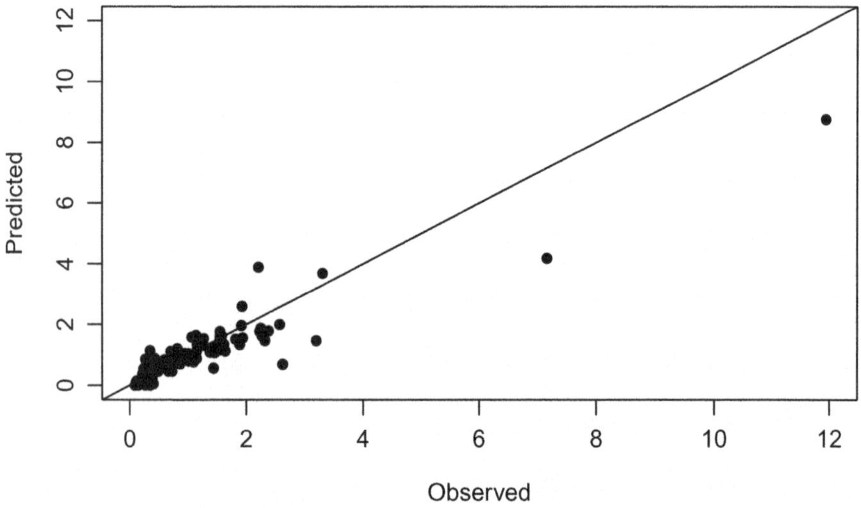

Fig. 10.16 Scatterplot of observed against predicted values of total carbon. Predictions are made by the cubist model on the calibration dataset for the example

We can evaluate the prediction using the evaluation measures.

```
soilspec::eval(datsoilspcV$TotalCarbon, soilvCubistPred, obj = "quant")
```

```
##         ME RMSE   r2   R2 rhoC  RPD RPIQ
## 1 -0.11 0.62 0.79 0.82 0.89 2.38 1.94
```

```
# predict the values from spectra at different moisture content
soilvCubistPredDry <- predict(epoCubistModel, Z0)
soilvCubistPredWet <- predict(epoCubistModel, Z1)
soilvCubistPredWet2 <- predict(epoCubistModel, Z2)

par(mfrow = c(1, 3))

# plot the prediction on low soil moisture content spectra
plot(soilC$TotalC, soilvCubistPredDry,
     xlab = "Observed",
     ylab = "Predicted",
     main = "Moisture = 5%",
     xlim = c(0, 12),
     ylim = c(0, 12),
     pch = 16)
abline(0, 1)

# plot the prediction on moderate soil moisture content spectra
plot(soilC$TotalC, soilvCubistPredWet2,
     xlab = "Observed",
     ylab = "Predicted",
     main = "Moisture = 9%",
     xlim = c(0, 12),
     ylim = c(0, 12),
     pch = 16)
abline(0, 1)

# plot the prediction on high soil moisture content spectra
plot(soilC$TotalC, soilvCubistPredWet,
     xlab = "Observed",
     ylab = "Predicted",
     main = "Moisture = 12%",
     xlim = c(0, 12),
     ylim = c(0, 12),
     pch = 16)
abline(0, 1)
```

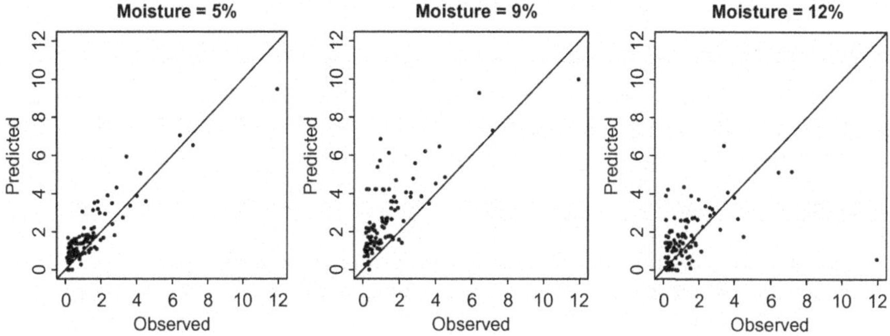

Fig. 10.17 Scatterplot of predicted against observed values of total carbon. Predictions are made using a cubist model on three varying levels of soil moisture from low (left) to high (right) soil moisture content. Spectra are corrected for soil moisture

Figure 10.17 shows the prediction on spectra corrected for moisture, to be compared to Fig. 10.13.

We also derive the accuracy measures (Sect. 9.1) between predicted and observed values of the total carbon for the three datasets with varying levels of moisture.

```
soilspec::eval(soilC$TotalC, soilvCubistPredDry, obj = "quant")
```

```
##     ME RMSE   r2   R2 rhoC  RPD RPIQ
## 1 0.4 0.82 0.64 0.83 0.91 2.43 1.55
```

```
soilspec::eval(soilC$TotalC, soilvCubistPredWet2, obj = "quant")
```

```
##      ME RMSE  r2   R2 rhoC  RPD RPIQ
## 1 1.29 1.83 0.5 0.15 0.66 1.09 0.69
```

```
soilspec::eval(soilC$TotalC, soilvCubistPredWet, obj = "quant")
```

```
##      ME RMSE   r2   R2 rhoC  RPD RPIQ
## 1 0.26 1.83 0.23 0.16 0.42 1.09 0.69
```

The accuracy measures show that removing the effect of moisture remarkably improves prediction accuracy.

In summary, the steps are:

1. EPO calibration
   - Calculate difference spectra **D** as the difference between moist and dry spectra.
   - Perform a principal component analysis (PCA) on **D** using the EPO function by defining c number of factors.
   - Obtain the projection matrix **P** from the `epo()` function.
2. Model calibration
   - Transform the spectra **X** from a library into the EPO space (**Z**) which is not affected by moisture, $\mathbf{Z} = \mathbf{XP}$.
   - Calibrate a model using the transformed spectra **Z** to predict a soil property.
3. Model prediction
   - Transform the spectrum of an unknown sample **x**: $\mathbf{z} = \mathbf{xP}$.
   - Use **z** in the model to predict a soil property based on a calibrated model.

The only parameter that needs to be chosen in EPO is `npc`, the number of dimensions of the principal components, which represents the unwanted part that needs to be removed. One way of determining `npc` is by cross-validation of a calibration model based on the transformed spectra. This means that `npc` may vary between different models for soil properties. Another solution is discussed in the next section using Wilks' Λ.

Determining number of components based on Wilks' Λ

Roger et al. (2003) suggested a simple way of determining the number of components `npc` by calculating how similar the transformed spectra of two or more moisture content compared to between sample spectra. This can be measured using a cluster separation metric called Wilks' Λ, which is the ratio between inter-group variance and the total variance (Webster 1971).

In this case, $\Lambda = \text{tr}(\mathbf{B})/\text{tr}(\mathbf{T})$, where tr(.) is the trace, **B** is the inter-group variance-covariance matrix of the aggregated transformed spectra (i.e. EPO transformed spectra averaged across all moisture levels) and **T** is the variance-covariance matrix of the EPO transformed spectra.

Before EPO transformation, the spectra of a sample at different moisture contents (intra-sample variation) differ more than the spectra of two different samples (inter-sample variation). After EPO transformation, different samples should be well separated in the spectral space (Wijewardane et al. 2016). Wilks' $\Lambda = 1$ indicates a perfect separation of samples, while a value of 0 indicates no separation.

The following code calculates Wilks' Λ as a function of number of components. We assign `npc = 0` for no transformation (Fig. 10.18).

```
D = as.matrix(spectra0Snv - spectra1Snv)
SC <- (spectra0Snv + spectra1Snv)/2
B <- cov(SC)
TrB <- sum(diag(B))
T <- cov(spectra1Snv)

nc <- 20
xc <- seq(0:nc)
wilks <- matrix(0, nrow = nc + 1, ncol = 1) # create

# no transformation
wilks[1] <- TrB/sum(diag(T))

# make a for loop to estimate the Wilks" lambda
for (i in 1:nc) {
    npc <- i
    P <- epo(D, npc)
    Z0 <- as.matrix(spectra0Snv) %*% P
    Z1 <- as.matrix(spectra1Snv) %*% P
    T <- cov(Z1)

    SC <- (Z0+Z1)/2
    B <- cov(SC)
    TrB <-sum(diag(B))

    wilks[i+1] <- TrB/sum(diag(T))
    }
```

```
# plot the Wilks" lambdas against number of component
plot(xc, wilks,
     type = "l",
     xlab = "Number of components",
     ylab = expression(paste("Wilks", ~Lambda)))
```

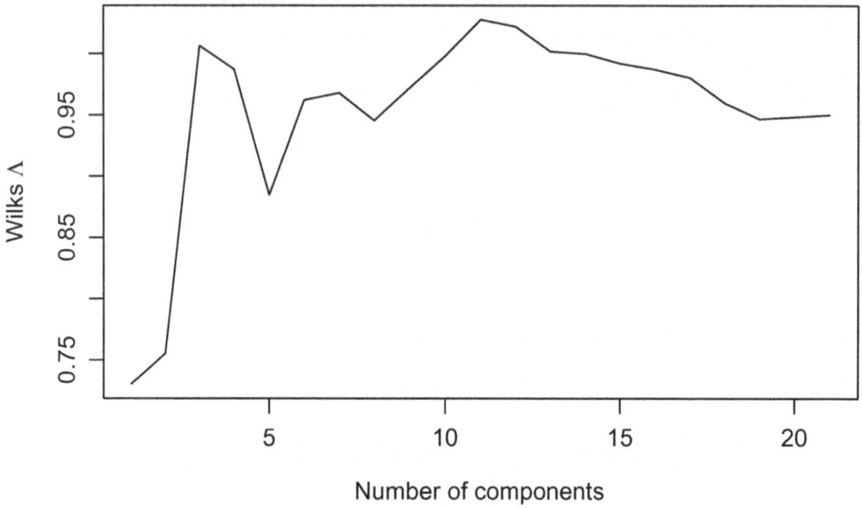

Fig. 10.18 Number of principal components used against values of Wilks' Λ

The first maximum of Wilks' Λ is at `npc = 3`. The three components that we used previously seemed a good choice.

Note that the DS and PDS methods can also be used to correct for other external factors influencing the spectra such as soil moisture. However, it is not recommended as the moisture content of soil in the field can be variable and not just at one moisture content. The DS method was shown efficient to correct for moisture in paddy soil in the field (which we assume to be always wet) by Ji et al. (2015).

References

Ben Dor E, Ong C, Lau IC (2015) Reflectance measurements of soils in the laboratory: standards and protocols. Geoderma 245–246:112–124

Ji W, Viscarra-Rossel RA, Shi Z (2015) Accounting for the effects of water and the environment on proximally sensed vis–NIR soil spectra and their calibrations. Eur J Soil Sci 66:555–565

Minasny B, McBratney AB, Bellon-Maurel V, Roger J-M, Gobrecht A, Ferrand L, Joalland S (2011) Removing the effect of soil moisture from NIR diffuse reflectance spectra for the prediction of soil organic carbon. Geoderma 167:118–124

Pimstein A, Notesco G, Ben-Dor E (2011) Performance of three identical spectrometers in retrieving soil reflectance under laboratory conditions. Soil Sci Soc Am J 75:746–759

Roger J-M, Chauchard F, Bellon-Maurel V (2003) EPO–PLS external parameter orthogonalisation of PLS application to temperature-independent measurement of sugar content of intact fruits. Chemom Intell Lab Syst 66:191–204

Viscarra-Rossel RA, Behrens T, Ben-Dor E, Brown D, Demattê J, Shepherd KD, Shi Z, Stenberg B, Stevens A, Adamchuk V, others (2016) A global spectral library to characterize the world's soil. Earth-Sci Rev 155:198–230

Wall ME, Rechtsteiner A, Rocha LM (2003) Singular value decomposition and principal component analysis. In: A practical approach to microarray data analysis. Springer, Boston, pp 91–109

Wang Y, Veltkamp DJ, Kowalski BR (1991) Multivariate instrument standardization. Anal Chem 63:2750–2756

Webster R (1971) Wilks's criterion: a measure for comparing the value of general purpose soil classifications. J Soil Sci 22:254–260

Wijewardane NK, Ge Y, Morgan CLS (2016) Moisture insensitive prediction of soil properties from VNIR reflectance spectra based on external parameter orthogonalization. Geoderma 267:92–101

Willis DE (1972) Internal standard method calculations. Chromatographia 5:42–43

The manufacturer's authorised representative in the EU is Springer Nature Customer Service Centre GmbH, Europaplatz 3, 69115 Heidelberg, Germany. If you have any concerns regarding our products, please contact ProductSafety@springernature.com

Printed and bound by CPI Group (UK) Ltd, Croydon, CR0 4YY
30/09/2026
02225695-0005